THE DISCOVERY OF GLOBAL WARMING

NEW HISTORIES OF SCIENCE, TECHNOLOGY,
AND MEDICINE

SERIES EDITORS

Margaret C. Jacob, Spencer R. Weart, and Harold J. Cook

SPENCER R. WEART

THE DISCOVERY OF

GLOBAL
WARMING

HARVARD UNIVERSITY PRESS

CAMBRIDGE, MASSACHUSETTS

LONDON, ENGLAND

2003

Library of Congress Cataloging-in-Publication Data
Weart, Spencer R., 1942–
The discovery of global warming / Spencer R. Weart.
p. cm. — (New histories of science, technology, and medicine)
ISBN 0-674-01157-0 (alk. paper)
1. Global warming—History.   2. Climatology—History.
I. Title.   II. Series.
QC981.8 G56W43 2003
551.6—dc21   2003040703

CONTENTS

One day as I was walking home after hours spent studying scientific papers on the possibility of climate change, I noticed the elegant maples lining my street, and wondered if they were near the southern end of their natural range. All at once, in my mind's eye I saw the maples dead—felled by global warming.

This book is a history of how scientists came to imagine such things: the history of the science of climate change. It is a hopeful book. It tells how a few people, through ingenuity, stubborn persistence, and a bit of luck, came to understand a grave problem even before any effects became manifest. And it tells how many other people, defying the old human habit of procrastinating until a situation becomes unbearable, began working out solutions. For there are indeed ways to keep global warming within tolerable bounds with a reasonable effort. It turns out that the trees on my street are red maples, a hardy species that will hold up if we choose the proper actions over the next decades.

The future actions we might take are not my subject. This book is a history of how we got to (and got to understand) our present situation. The long struggle to grasp how humanity could be changing the weather was an obscure effort. For many decades it was pursued by only a few individuals, scarcely known except to their immediate colleagues. Yet their stories may be as important for the future of our civilization as any history of politics, wars, and social upheavals.

If you discover that termites have infiltrated your house and your roof is about to fall in, you know you must act. The discovery of global warming was never so clear. In 1896, a lonely Swedish scientist discovered global warming—as a theoretical concept, which most other experts declared implausible. In the 1950s, a few scientists in California discovered global warming—as a possibility, a risk that might perhaps come to pass in a remote future. In 2001, an extraordinary organization mobilizing thousands of scientists around the world discovered global warming—as a phenomenon that had measurably begun to affect the weather and was liable to get much worse. That was when we got the report from the termite inspector. But it was only the top item in a tall and messy stack, a record of so much uncertainty and confusion that most people who have heard about the issue still cannot say what, if anything, they should do.

We have hard decisions to make. Our response to the threat of global warming will affect our personal well-being, the evolution of human society, indeed all life on our planet. One aim of this book is to help the reader understand our predicament by explaining how we got here. By following how scientists in the past fought their way through the uncertainties of climate change, we can be better prepared to judge why they speak as they do today. More, we can understand better how scientists address the many other questions where they have an important voice.

How do scientists reach reliable conclusions? Our familiar picture of discovery, learned from the old core sciences like physics or biology, shows an orderly parade of observations and ideas and experiments. We like to think it ends with an answer, a clear statement about what we can do. Such a logical sequence, with definitive results, does not describe work in interdisciplinary fields like the study of climate change (actually, it often doesn't describe the old core sciences either). The story of the discovery of global warming looks less like a processional march than like a scattering of groups wandering around an immense landscape. Thousands of people are laboring on studies that may tell something about climate change only by chance. Many of the scientists are scarcely aware of one another's existence. Over here we find a computer master calculating the flow of glaciers, over there an experimenter rotating a dishpan of water on a turntable, and off to the side a student with a needle teasing tiny shells out of a lump of mud. This kind of science, where specialties are only partly in contact, has become widespread as scientists labor to understand increasingly complex topics.

The tangled nature of climate research reflects nature itself. The Earth's climate system is so irreducibly complicated that we will never grasp it completely, in the way that one might grasp a law of physics. These uncertainties infect the relationship between climate science and policy-making. Debates about climate change can get as confusing as arguments over the social consequences of welfare payments. To deal with this problem, climate scientists created remarkable new policy mechanisms. Describing such connections between science and society at large is another aim of this book. By tracing how scientists, politicians, journalists, and ordinary citizens pushed and pulled at one another in the past, we can be better prepared to deal with the fatal issues that confront us.

Climate change is not one story but many parallel stories, only sporadically connected. This book artificially weaves them together into a single overview. For those who wish to look deeper, I offer a supplementary Web site containing more than two dozen essays running in parallel, interconnected by over 700 hyperlinks. The Web site contains three times as much material as this book, filling in important historical and technical details and offering extended case studies of characteristic aspects. The Web site also has references to well over 1,000 scientific and historical publications and links not noted in this book. To get a better feeling for the historical shape of the discovery of global warming, and to explore particular topics more fully, visit **http://www.aip.org/history/climate.**

This work was supported by the American Institute of Physics and grants from the National Science Foundation's Program in Science & Technology Studies and the Alfred P. Sloan Foundation. I am grateful for the indispensable help provided by scientists and historians who generously gave me interviews, comments (sometimes extensive), or access to documents, including A. Arakawa, W. Broecker, K. Bryan, R. A. Bryson, R. Charlson, J. Eddy, P. Edwards, T. Feldman, J. Fleagle, J. Fleming, J. Hansen, C. D. Keeling, S. Manabe, J. Smagorinsky and R. M. White.

THE DISCOVERY OF GLOBAL WARMING

CHAPTER ONE

HOW COULD CLIMATE CHANGE?

People have always talked about unusual spells of weather, but in the 1930s the talk took an unusual turn. Old folks began to insist that the weather truly wasn't what it used to be. The daunting blizzards they remembered from their childhoods back in the 1890s, the freezing up of lakes in early fall, all that had ended—the younger generation had it easy. The popular press began to run articles that claimed winters really had gotten milder. Meteorologists scrutinized their records and confirmed it: a warming trend was under way. Experts told science reporters that frosts were coming later, and that wheat and codfish could now be harvested in northern zones where they had not been seen for centuries. As *Time* magazine put it in 1939, "gaffers who claim that winters were harder when they were boys are quite right . . . weather men have no doubt that the world at least for the time being is growing warmer."[1]

Nobody worried about the change. The meteorologists explained that weather patterns always did vary modestly, in cycles lasting a few decades or centuries. If the mid-twentieth century happened to be a time of warming, so much the better. A typical popular article of 1950 promised that "vast new food-producing areas will be put under cultivation." To be sure, if the warming continued, new deserts might appear. And the oceans might rise to flood coastal cities—"another deluge, such as the catastrophe recorded in the Bible."[2] All that was plainly just colorful speculation about a remote future. Many professional meteorologists doubted that there was in fact any worldwide warming trend. All they saw were normal, temporary, regional fluctuations. And if there was global warming, a magazine report explained that "Meteorologists do not know whether the present warm trend is likely to last 20 years or 20,000 years."[3] In 1952 (Aug. 10) the *New York Times* remarked that thirty years hence, people might look back fondly on the mild winters of the 1950s.

One man challenged the consensus of the experts. In 1938 Guy Stewart Callendar had the audacity to stand before the Royal Meteorological Society in London to talk about climate. Callendar was out of place, for he was no professional meteorologist, not even a scientist, but an engineer who worked on steam power. He had an amateur interest in climate and had spent many hours of spare time putting together weather statistics as a hobby. He had confirmed (more thoroughly than anyone else) that the numbers indeed showed global warming. Now Callendar told the meteorologists he knew what was responsible. It was us, human industry. Everywhere we burned fossil fuels we emitted millions of tons of carbon dioxide gas (CO_2), and that was changing the climate.[4]

This idea was not new, for the basic physics had been worked out during the nineteenth century. Early in the century, the French scientist Joseph Fourier had asked himself a question. It was a deceptively simple question, of a sort that physics theory was just then beginning to learn how to attack: what determines the average temperature of a planet like the Earth? When light from the Sun strikes the Earth's surface and warms it up, why doesn't the planet keep heating up until it is as hot as the Sun itself? Fourier's answer was that the heated surface emits invisible infrared radiation, which carries the heat energy away into space. But when he calculated the effect with his new theoretical tools, he got a temperature well below freezing, much colder than the actual Earth.

The difference, Fourier recognized, was due to the Earth's atmosphere, which somehow keeps some of the heat radiation in. He tried to explain this by comparing the Earth with its covering of air to a box covered with a pane of glass. The box's interior warms up when sunlight enters, while the heat cannot escape. The explanation sounded plausible, and by Callendar's time a few scientists had begun to speak of a "greenhouse effect" that keeps the Earth from freezing. It is a misnomer, for real greenhouses stay warm for their own reasons (the main effect of the glass is to keep the air, heated by sun-warmed surfaces, from wafting away). As Fourier recognized, the way the atmosphere holds in heat on the entire Earth is more subtle. The atmosphere's trick is to somehow intercept a part of the infrared radiation emitted from the surface, preventing it from escaping into space.

The correct reasoning was first explained lucidly by a British scientist, John Tyndall. Tyndall pondered how the atmosphere might control the earth's temperature, but he was stymied by the opinion, held by most scientists at the time, that all gases are transparent to infrared radiation. In 1859 he decided to check this out in his laboratory. He confirmed that the main gases in the atmosphere, oxygen and nitrogen, are indeed transparent. He was ready to quit when he thought to try coal gas. This was an industrial gas produced by heating coal, mostly methane, which was used for lighting. It was right at hand, piped into his laboratory. He found that for heat rays, this gas was as opaque as a plank of wood. Thus the Industrial Revolution, intruding into Tyndall's laboratory in the form of a gas jet, declared its significance for the planet's heat balance. Tyndall went on to try other gases, and found that the gas CO_2 was likewise opaque—what we would now call a greenhouse gas.

A bit of CO_2 is found in the Earth's atmosphere, and although it is only a few parts in ten thousand, Tyndall saw how it could bring warming. A fraction of the infrared radiation rising from the surface is absorbed by CO_2 in the middle levels of the atmosphere.

Its heat energy is transferred into the air itself rather than escaping into space. Not only is the air warmed, but also some of the energy trapped in the atmosphere is radiated back to the surface and warms it. Thus the temperature of the Earth is maintained at a higher level than it would be without the CO_2. Tyndall put it neatly: "As a dam built across a river causes a local deepening of the stream, so our atmosphere, thrown as a barrier across the terrestrial [infrared] rays, produces a local heightening of the temperature at the Earth's surface."[5]

Tyndall's interest in all this had begun in a wholly different type of science. He hoped to solve a puzzle that was exciting great controversy among the scientists of his day: the prehistoric Ice Age. The claims were incredible, yet the evidence was eloquent. The scraped-down rock beds, the bizarre deposits of gravel found all around northern Europe and the northern United States, these looked exactly like the effects of Alpine glaciers, only immensely larger. Amid fierce debate, scientists were coming to accept the incredible. Long ago—although not so long as geological time went, for Stone Age humans had lived through it—northern regions had been buried a mile deep in continental sheets of ice. What could have caused this?

Changes in the atmosphere were one possibility, although not a promising one. Of the atmospheric gases, CO_2 was not an obvious suspect, since there is so little of it in the atmosphere. The really important "greenhouse" gas is H_2O, simple water vapor. Tyndall found that it readily blocks infrared radiation. He explained that water vapor "is a blanket more necessary to the vegetable life of England than clothing is to man. Remove for a single summer-night the aqueous vapor from the air . . . and the sun would rise upon an island held fast in the iron grip of frost."[6] So if something dried out the atmosphere, that might cause an ice age. At present, the atmosphere's average humidity seemed to be maintained in some sort of automatic balance, in tandem with the global temperature.

The riddle of the ice age was taken up in 1896 by a scientist in Stockholm, Svante Arrhenius. Suppose, he said, the amount of CO_2 in the atmosphere were changed. For example, a spate of volcanic eruptions might spew out vast quantities of the gas. This would raise the temperature a bit, and that small increment would have an important consequence: the warmer air would hold more moisture. Because water vapor is the truly potent greenhouse gas, the additional humidity would greatly enhance the warming. Conversely, if all volcanic emissions happened to shut down, eventually the CO_2 would be absorbed into soil and ocean water. The cooling air would hold less water vapor. Perhaps the process would spiral into an ice age.

Cooling that causes less water vapor in the air that causes more cooling that causes . . . this is the kind of self-reinforcing cycle that today we call "feedback." The concept was both elementary and subtle—easy to grasp, but only after somebody pointed it out. In Arrhenius's day only a few insightful scientists understood that such effects could be crucial for understanding climate. The first important example had been worked out in the 1870s by a British geologist, James Croll, as he pondered possible causes of the ice age. When snow and ice had covered a region, he noted, they would reflect most of the sunlight back into space. Bare, dark soil and trees would be warmed by the Sun, but a snowy region would tend to remain cool. If India were somehow covered with ice, he said, its summers would be colder than England's. Croll further argued that when a region cooled down, the pattern of winds would change, which would in turn change ocean currents, perhaps removing more heat from the region. Once something started an ice age, the pattern could well become self-sustaining.

Such complex effects were far beyond anyone's ability to calculate at that time. The most Arrhenius could do was to estimate the immediate effects of changing the level of CO_2. But he realized that he could not overlook the crucial changes in water vapor as the temperature rose or fell. He would have to figure humidity into his calculations. This was far more than anyone else had attempted.

The numerical computations cost Arrhenius month after month of tedious pencil work. He calculated the atmospheric moisture and the radiation entering and leaving the Earth for each zone of latitude. It seems he undertook the massive task partly as an escape from melancholy: he had just been through a divorce, losing not only his wife but custody of their little boy. The countless computations could hardly be justified scientifically. Arrhenius had to overlook many features of the real world, and the data he used for how gases absorbed radiation were far from reliable. Nevertheless he came up with numbers that he published with some confidence. If he was far from proving how the climate *would* change if CO_2 varied, he did in truth get a rough idea of how it *could* change. He announced that cutting the amount of CO_2 in the air by half would cool the world by maybe 5°C (that is, 8° Fahrenheit). That might not seem like a lot. But thanks to feedbacks, as extra snow accumulated and reflected sunlight, it might be enough to bring on an ice age.

Were such large changes in atmospheric composition possible? For that question Arrhenius turned to a colleague, Arvid Högbom. Högbom had compiled estimates for how CO_2 cycles through natural geochemical processes—emission from volcanoes and uptake by the oceans and so forth—and he had come up with a strange new thought. It had occurred to him to calculate the amounts of CO_2 emitted by factories and other industrial sources. Surprisingly, he found that the rate at which human activities were adding the gas to the atmosphere was roughly the same as the rates at which natural processes emitted and absorbed the gas. The added gas was not much compared with the volume of CO_2 already in the atmosphere—the amount released from the burning of coal in the year 1896 would raise the level by scarcely a thousandth part. But the additions might matter if they continued long enough. Arrhenius calculated that doubling the CO_2 in the atmosphere would raise the Earth's temperature some 5 or 6°C.

The idea of humans massively perturbing the atmosphere did not trouble Arrhenius. It was not just that warming seemed like a good thing in chilly Sweden. Arrhenius, like nearly everyone at the end of the nineteenth century, expected any technological change would be for the best. People believed that scientists and engineers would solve all the problems of poverty in the centuries to come. They would turn deserts into gardens! In any case, Arrhenius figured it would take a couple of thousand years to double the amount of CO_2 in the air. In his day hardly anyone grasped how the planet's population was doubling and redoubling. Still less did people understand how the use of resources was mounting even more swiftly than populations. Barely a billion people populated the world, mostly ignorant peasants living like medieval serfs. It scarcely seemed reasonable to imagine that humans could change the entire planet's atmosphere, unless perhaps in some remote and fantastic future. Arrhenius had not quite discovered global warming, but only a curious theoretical concept.

Even as abstract theory, there were scientific reasons to dismiss Arrhenius's idea. Most telling was a simple laboratory measurement that seemed to refute the entire principle of greenhouse warming. A few years after Arrhenius published his hypothesis, an experimenter sent infrared radiation through a tube filled with CO_2. He put in as much of the gas in total as would be found in a column of air reaching to the top of the atmosphere. The amount of radiation that got through the tube scarcely changed when he doubled the quantity of gas. The reason was that CO_2 absorbs radiation only in specific bands of the spectrum. It took only a trace of the gas to produce bands that were "saturated"—so thoroughly opaque that adding more gas could make little difference. Moreover, water vapor already absorbed infrared radiation in the same region of the spectrum. Evidently, the planet already had the maximum possible greenhouse effect. By 1910 most scientists thought Arrhenius's calculation was altogether wrong.

To kill any lingering doubts, other scientists pointed out a still more fundamental objection. They held that it was impossible for CO_2 to build up in the atmosphere at all. The atmosphere is only a wisp that contains little of the material on the Earth's surface, by comparison with the huge quantities locked up in minerals and in the oceans. For every molecule of CO_2 in the air, there are about 50 dissolved in seawater. If humanity added more of the gas to the air, nearly all of it would eventually wind up in the oceans.

Furthermore, scientists saw that Arrhenius had grossly oversimplified the climate system in his calculations. He had ignored such things as the way wind patterns and ocean currents might change as the temperature changed. And he had skipped over even more elementary effects. For example, if more water vapor was held in the air as the Earth got warmer, surely the moisture would make more clouds. The clouds would reflect sunlight back into space before the energy ever reached the surface, and so the Earth should hardly warm up after all.

These objections conformed to a view of the natural world that was so widespread that most people thought of it as plain common sense. In this view, the way cloudiness rose or fell to stabilize temperature, or the way the oceans maintained a fixed level of gases in the atmosphere, were examples of a universal principle: the Balance of Nature. Hardly anyone imagined that human actions, so puny among the vast natural powers, could upset the balance that governed the planet as a whole. This view of Nature—suprahuman, benevolent, and inherently stable—lay deep in most human cultures. It was traditionally tied up with a religious faith in the God-given order of the universe, a flawless and imperturbable harmony. Such was the public belief, and scientists are members of the public, sharing most of the assumptions of their culture. Once scientists found plausible arguments explaining that the atmosphere and climate would remain unchanged within a human timescale—just as everyone expected—they stopped looking for possible counterarguments.

Of course everyone knew climate could vary. From the old folks' tales of the great blizzards of their childhood to the devastating Dust Bowl drought of the 1930s, ideas about climate included a dose of catastrophe. But a catastrophe was (by definition) something transient, with things reverting to normal after a few years. A few scientists speculated about greater climate shifts. For example, had a waning of rainfall over slow centuries caused the downfall of ancient Near Eastern civilizations? Most doubted it. And if such changes really did happen, everyone assumed they randomly struck one or another local region, not the entire planet.

To be sure, everyone knew there had been vast global climate changes in the distant past. Geologists were mapping out the ice age—or rather ice ages. For it turned out that the tremendous sheets of ice had ground halfway down America and Europe and back not once, but over and over again. Looking still farther in the past, geologists found a tropical age when dinosaurs basked in regions that were now arctic. A popular theory suggested that the dinosaurs had perished when the Earth cooled over millions of years—climate change could be serious if you waited long enough. The most recent ice age likewise had come to a gradual end, geologists reported, as the Earth returned to its present temperature over tens of thousands of years. If a new ice age was coming, it should take as long to arrive.

The rate of advance and retreat of the great ice sheets had been no faster than present-day mountain glaciers were seen to move. That fitted nicely with "the uniformitarian principle." This principle held that the forces that molded ice, rock, sea, and air did not vary over time, or, as some put it, nothing could change otherwise than the way things were seen to change in the present. The principle was cherished by geologists as the very foundation of their science, for how could you study anything scientifically unless the rules stayed the same? The idea had taken hold during a century of disputes. Scientists had painfully given up traditions that explained certain geological features by Noah's Flood or other abrupt supernatural interventions. The passionate debates between "uniformitarian" and "catastrophist" theories had only partly brought science into conflict with religion. Many pious scientists and rational preachers could agree that everything happened by natural processes in a world governed by a reliable God-given order.

Ideals of consistency pervaded not only the study of climate, but also the careers of those who studied it. Through the first half of the twentieth century, climate science was a sleepy backwater. People who called themselves "climatologists" were mostly drudges who kept track of average seasonal temperatures, rainfall, and the like. Typical were the workers at the U.S. Weather Bureau, "the stuffiest outfit you've ever seen," as a member of a later generation of research-oriented geophysicists put it.[7] Their job was to compile statistics on past weather, in order to advise farmers what crops to grow or tell engineers how great a flood was likely over the lifetime of a bridge. These climatologists' products were highly appreciated by their customers (such studies continue to this day). And their tedious, painstaking style of scientific work would turn out to be indispensable for studies of climate change. Yet the value of this kind of climatology to society was based on the conviction that statistics of the past half-century or so could reliably describe conditions for many decades ahead. Textbooks started out by describing the term "climate" as a set of weather data averaged over temporary ups and downs—it was stable *by definition*.

The few who went beyond statistics to attempt explanations used only the most elementary physics. The temperature and precipitation of a region were set by the amount of sunlight at that latitude, the prevailing winds, the location of ocean currents that might warm the winds or mountain ranges that might block them, and the like. As late as 1950, if you looked in a university for a climatologist, you might find one in the geography department, but not in a department of atmospheric sciences or geophysics (hardly any such departments existed anyway). The field was rightly regarded, as one practitioner complained, as "the dullest branch of meteorology."[8]

Nevertheless, plenty of vigorous speculation about climate change was heard, less from professional climatologists than from outsiders like Callendar. By the time he addressed the Royal Meteorological Society, the meteorologists had already heard only too many gaudy ideas. For although the ice ages lay in the remote past, seemingly of no practical concern, they loomed as a grand intellectual challenge. It was not the vague possibility of global warming that intrigued the few people who thought about climate change, but the stupendous advance and retreat of continental ice sheets. Tyndall, Arrhenius, Callendar, and not a few others hoped to win lasting fame by solving that notorious puzzle. From time to time newspapers would amuse their readers and embarrass climatologists by writing up some halfway plausible theory announced by one or another university professor or eccentric amateur. As one writer put it, "Everyone has his own theory—and each sounds good—until the next lad comes along with his theory and knocks the others into smithereens."[9]

When announcing theories, the professional scientists were not always easy to distinguish from the amateurs. Climatology could hardly be scientific when meteorology itself was more art than science. The best attempts to use physics and mathematics to describe weather—or even simple, regular features of the planet's atmosphere like the trade winds—had gotten nowhere. Much as climatologists could only try to predict a season by looking at the record of previous years, so meteorologists could only try to predict the next day's weather by comparison with weather of the past. Sometimes this was done systematically, matching the current weather map to an atlas of old weather maps, but more often a forecaster just looked at the current situation and drew on experience with a combination of simple calculations, rules of thumb, and personal intuition. A canny amateur with no academic credentials could predict rain as successfully as a Ph.D. meteorologist. Indeed, through the first half of the twentieth century most of the "professionals" in the U.S. Weather Bureau lacked any college degree.

Yet it is the nature of scientists never to cease trying to explain things. If there was no accepted theory—indeed the very word "theory" brought a skeptical frown from climatologists—there was a list of forces that could plausibly shift climate. Scientists turned up candidates everywhere, from the interior of the Earth to outer space. The possibilities spanned half a dozen different sciences.

First in line was geology. The most widely accepted explanations for the ice ages looked to the Earth's interior. If some great upheaval raised mountain ranges to block prevailing winds, for example, climates would surely change. Similarly, the raising or lowering of an island chain might change the course of the Gulf Stream so that its warmth would not reach Europe. Such forces could perhaps explain the difference between a warm age of dinosaurs and an era of ice ages. However, mountain-building took millions of years, whereas the continental ice sheets had surged back and forth in mere hundreds of thousands. To explain such relatively rapid climate shifts, geologists would have to look for other forces.

In 1783 a volcanic fissure in Iceland erupted with enormous force, pouring out cubic kilometers of lava. Layers of ash and cinders snowed down upon the island. The grass died and three-quarters of the livestock starved to death, followed by a quarter of the people. A peculiar haze dimmed the sunlight over western Europe for months. Benjamin Franklin, visiting France, noticed the unusual cold that summer, and speculated that it might have been caused by the volcanic "fog." The idea caught on. By the end of the nineteenth century, most scientists believed that volcanic eruptions might indeed affect large regions, even the entire planet. Perhaps smoky skies during spells of massive volcanic eruptions were the cause of each advance of glaciers during the ice ages.

Other scientists suggested that the answer did not lie in geology, but in the oceans. The huge bulk of the oceans contains the main ingredients of climate: far more water than the tenuous atmosphere, of course, plus most of the planet's gases, dissolved in the seawater. And just the top few meters of the oceans hold more heat energy than the entire atmosphere. The chief feature of the Earth's surface heat circulation was discovered in the nineteenth century, when water hauled up from the deeps was found to be nearly freezing everywhere in the world (oceanographers claim the investigation began when a scientist traveling on a steamship in the tropics saw a steward chilling bottles of wine by dunking them overboard). This water must have sunk in arctic regions and flowed equatorwards along the bottom. The idea made sense, since water would be expected to sink where arctic winds made it colder, and therefore denser.

However, the warm tropical seas rapidly evaporated moisture. The moisture eventually came down as rain and snow farther north, leaving the equatorial waters more salty. When water gets more salty it gets denser, so shouldn't ocean waters sink in the tropics? Around the turn of the century a versatile American scientist, T. C. Chamberlin, took an interest in the question. He figured that "the battle between temperature and salinity is a close one . . . no profound change is necessary to turn the balance."[10] Perhaps in earlier geological eras, when the poles had been warmer, salty ocean waters had plunged in the tropics and come up near the poles. This reversal of the present circulation, he speculated, could have helped maintain the uniform global warmth seen in the distant past.

This was a subtle sort of explanation, and few took it up. The circulation of the oceans, like much else, was pictured as a placid equilibrium, perpetually following the same track. That was what scientists observed, if only because measurements at sea were few and difficult. Nobody had seen any reason to make the long effort that would be needed to develop a technology for making precise measurements of the seas. Oceanographers traced currents by the simple expedient of throwing bottles into the ocean. Such measurements could not detect a change in the pattern of currents even if the scientists had thought to look for it. Gradually the oceanographers sketched out a pattern of stable currents.

Harald Sverdrup displayed the full picture in a textbook published in 1942. He described, as one item in a list of many ocean features, how cold, dense water sinks near Iceland and Greenland and flows southward in the deeps. To complete the North Atlantic cycle, warm water from the tropics drifts slowly northward near the surface. Winds presumably add a push, although their effect was uncertain. Oceanographers debated the relative importance of trade winds, heat, and salinity in the North Atlantic circulation, without having any means to resolve the question. It seemed a minor issue anyway. Sverdrup did not remark that the immense volume of warm water drifting northward might be significant for climate. Like all oceanographers of his time, he gave most of his attention to rapid surface currents like the Gulf Stream. Only those seemed to count for what really concerned oceanographers—navigation, fisheries, and regional climates.

Another idea about causes of climate change came from an entirely different direction. Since the ancient Greeks, ordinary folk and scholars alike had wondered whether chopping down a forest or grazing a prairie to bare dirt might change the weather in the vicinity. It seemed common sense that shifting the plant cover from trees to wheat or from grass to desert would affect temperature and rainfall. Americans in the nineteenth century argued that settlement of the country had brought a less savage climate, and sodbusters who moved into the Great Plains boasted that "rain follows the plough."

By the end of the nineteenth century, meteorologists had accumulated enough reliable weather records to test the idea. It failed the test. Even the transformation of the entire ecosystem of eastern North America from woods to farmland had made little evident difference to the climate. Apparently the atmosphere was indifferent to biology.[11] That seemed reasonable enough. Whatever forces could change climate were surely far mightier than the thin scum of organic matter that covered some patches of the planet's surface.

A bare handful of scientists thought otherwise. The deepest thinker was the Russian geochemist Vladimir Vernadsky. From his work mobilizing industrial production for the First World War, Vernadsky recognized that the volume of materials produced by human industry was approaching geological proportions. Analyzing biochemical processes, he concluded that the oxygen, nitrogen, and CO_2 that make up the Earth's atmosphere are put there largely by living creatures. In the 1920s he published works arguing that living organisms constituted a force for reshaping the planet comparable to any physical force. Beyond this he saw a new and still greater force coming into play: intelligence. Vernadsky's visionary pronouncements about humanity as a geological force were not widely read, however, and struck most readers as nothing but romantic ramblings.

A stronger claim to explain climate came from the seemingly most unworldly of sciences, astronomy. It began with a leading eighteenth-century astronomer, William Herschel. He noted that some stars varied in brightness, and that our Sun is itself a star. Might the Sun vary its brightness, bringing cooler or warmer periods on Earth? Speculation increased in the mid-nineteenth century, following the discovery that the number of spots seen on the Sun rose and fell in a regular 11-year cycle. It appeared that the sunspots reflected some kind of storminess on the Sun's surface—violent ac-

tivity that had measurable effects on the Earth's magnetic field. Perhaps sunspots connected somehow with weather—with droughts,
for example? That would raise or lower the price of grain, so some
people searched for connections with the stock market. The study
of sunspots might give hints about longer-term climate shifts too.

Most persistent was Charles Greeley Abbot of the Smithsonian
Astrophysical Observatory. The Observatory already had a program of measuring the intensity of the Sun's radiation received at
the Earth, called the "solar constant." Abbot pursued the program
single-mindedly, and by the early 1920s he had concluded that
the solar constant was misnamed. His observations showed large
variations over periods of days, which he connected with sunspots
passing across the face of the Sun. Over a term of years the more ac-
tive Sun seemed brighter by nearly one percent. As early as 1913
Abbot had announced that he could see a plain correlation between
the sunspot cycle and cycles of temperature on Earth. Self-con-
fident and combative, Abbot defended his findings against all objections, meanwhile telling the public that solar studies would bring
wonderful improvements in weather prediction. Other scientists
were quietly skeptical, for the variations Abbot reported teetered at
the very edge of detectability.

The study of cycles was generally popular through the first half
of the century. Governments had collected a lot of weather data to
play with, and inevitably people found correlations between sunspot cycles and selected weather patterns. If rainfall in England
didn't fit the cycle, maybe storminess in New England would. Respected scientists and enthusiastic amateurs insisted they had
found patterns reliable enough to make predictions.

Sooner or later, though, every prediction failed. An example was
a highly credible forecast of a dry spell in Africa during the sunspot minimum of the early 1930s. When the period turned out
wet, a meteorologist later recalled, "the subject of sunspots and
weather relationships fell into disrepute, especially among British

meteorologists who witnessed the discomfiture of some of their most respected superiors." Even in the 1960s, he said, "For a young [climate] researcher to entertain any statement of sun-weather relationships was to brand oneself a crank."[12] Yet *something* had caused the ice ages. Long-term cycles of the Sun were as likely a possibility as any.

It seemed there was scarcely any science that could not make a claim on climate—even celestial mechanics. In the 1870s James Croll published calculations of how the gravitational pulls of the Sun, Moon, and planets subtly affect the Earth's motions. The inclination of the Earth's axis and the shape of its orbit around the Sun oscillate gently in cycles lasting tens of thousands or hundreds of thousands of years. During some millennia the Northern Hemisphere would get slightly less sunlight during the winter than it would get during other times. Snow would accumulate, and Croll argued this could bring on a self-sustaining ice age. The timing of such changes could be calculated exactly using classical mechanics (at least in principle, for the mathematics were thorny). Croll believed that the timing of the astronomical cycles roughly matched the timing of the ice ages.

The glacial periods did seem to follow a cyclical pattern. The advances and retreats of ancient glaciers could be detected from long mounds of gravel (moraines) that marked where the ice had halted, and in fossil shorelines of lakes in regions that were now dry. From meticulous studies of such surface features, first in Europe and then around the world, a generation of geologists constructed a sequence. They found four distinct advances and retreats, four ice ages. Croll's timing did not match this sequence at all.

Nevertheless a few enthusiasts pursued the theory. Taking the lead was a Serbian engineer, Milutin Milankovitch. Between the two world wars he not only improved the tedious calculations of the varying distances and angles of the Sun's radiation, but also came up with an important new idea. Suppose there was a particular time when the sunlight falling in a high-latitude zone of a given hemisphere was so weak, even in the summer, that the snow that fell in winter would not all melt away? It would build up, year after year. As Croll had pointed out, a covering of snow would reflect sunlight. So the snowfield might grow over centuries into a continental ice sheet.

By the 1940s, some climate textbooks were teaching that Milankovitch's calculations gave a plausible solution to the problem of timing the ice ages. Supporting evidence came from "varves," a Swedish word for layers of silt covering the bottom of northern lakes. A layer was laid down each year by the spring runoff. Scientists extracted samples of slick gray clay from lake beds and painstakingly counted the layers. Some researchers reported finding a 21,000-year cycle of changes. That approximately matched the timing for a swaying of the Earth's axis which Milankovitch had calculated (the "precession of the equinoxes").

But Milankovitch's numbers, like Croll's, failed to match the standard sequence of four ice ages found in every geological textbook. Worse, there was a basic physical argument against the whole theory. The variations that Milankovitch computed in the angle and intensity of incoming sunlight were slight. Most scientists thought it far-fetched to claim that a tiny shift in sunlight, too small to be noticed by the naked eye, could bury half a continent under ice. So what had caused the ice ages? That was still anybody's guess.

Thus when Callendar stood up before the Royal Meteorological Society in 1938, he was following many others who had speculated about climate change. Pointing to measurements of CO_2 he had dug up in old and obscure publications, he argued that the level of the gas in the atmosphere had risen a bit since the early nineteenth century. The experts were dubious. They understood that nobody had been able to make reliable measurements of the slight trace of CO_2 in the atmosphere. Callendar seemed to be picking only the data that supported his case (only in retrospect can we confirm that his judgment was pretty good). To be sure, Callendar had compiled the most convincing evidence yet that global temperatures had been rising. But was there any reason to connect the rise with CO_2?

It was not a pressing issue. Callendar himself thought global warming would be a good thing for humanity, helping crops to grow more abundantly. In any case, he calculated we would not raise the average global temperature much or soon, maybe one degree by the end of the twenty-second century. The meteorologists in his audience found it all intriguing but unconvincing, and dismissed him with a few condescending remarks.

So the debates continued. Some experts championed personal theories about *the* cause of climate change, the single dominant force. Most scientists gave short shrift to any theory whatsoever. They set aside climate change as a puzzle too difficult for anyone to solve with the tools at hand. The idea that humans were influencing global climate by emitting CO_2 sat on the shelf with the other bric-a-brac, a theory more peculiar and unattractive than most.

CHAPTER TWO

DISCOVERING A POSSIBILITY

Charles David Keeling—Dave to his friends—loved chemistry, and he loved the outdoors. As a postdoctoral student at the California Institute of Technology in the mid-1950s, he was committed to the sterile stinks of the laboratory, but he spent all the time he could spare traveling mountains and woodland rivers. He chose research topics that would keep him in direct contact with wild nature. Monitoring the level of CO_2 in the open air would do just that. Keeling's work was one example of how geophysics research often rested on love of the true world itself. On a lonely tundra or on a ship plowing the restless seas, when scientists devoted their years to research topics that many of their peers thought of minor import, part of the reason might be that these particular scientists could not bear to spend all their lives indoors. Yet their research sometimes turned out to be more significant than even they had hoped.

The study of atmospheric CO_2 had little to recommend itself to an ambitious scientist. Aside from the remote possibility that it might play some role in climate change over thousands of years, there was only a mild curiosity about how the winds carried around stuff that crops needed to grow, including the carbon in CO_2. A group in Scandinavia had attempted a monitoring program. Their measurements of CO_2 fluctuated widely from place to place, and even from day to day, as different air masses passed through carrying pulses of gas emitted by a forest or a factory. "It seems almost hopeless," one expert confessed, "to arrive at reliable estimates of the atmospheric carbon-dioxide reservoir and its secular changes by such measurements."[1] And regardless of Keeling's personal interest, would any agency give him money to make a new attempt?

That short question has a long and interesting answer. It begins with the revolutionary changes that the Second World War and the onset of the Cold War brought to the American scientific community. Consider, as a typical case, the transformation of meteorology. Generals and admirals, knowing well how battles can turn on the weather, needed meteorologists. The U.S. military turned to institutions like the University of Chicago, where a newly created department of meteorology was one of the few places in the world where the subject was being studied with full scientific rigor.

That was thanks to Carl-Gustav Rossby. Trained in mathematical physics in Stockholm, Rossby had come to the United States in 1925 to work in the Weather Bureau. He soon left the somnolent Bureau in disgust. Outstanding not only as a theorist but also as an entrepreneur and organizer, he created the nation's first professional meteorology program at the Massachusetts Institute of Technology. In 1942 he moved on to Chicago to establish another. Rossby was a leader of a scattered group of meteorologists who were determined to make the study of climate truly scientific. In place of the traditional climatology that merely listed descriptions of the "normal," unchanging climate in each geographical region, they would derive a more complex understanding of climate from basic principles of physics. The goal was an exercise in pure mathematics, deliberately remote from the fluctuations of actual weather and the uncertainties of daily predictions.

The scientific project was delayed by the war, but the Chicago meteorology department expanded tremendously during the war years. Rossby and his colleagues trained some 1,700 military meteorologists in one-year courses. Similarly, in other institutions and other fields of geophysics, teaching and research enterprises thrived as fighting men sought every possible scrap of knowledge about the winds, oceans, and beaches where they would do battle. This paid off as meteorologists and other geoscientists provided life-or-death information for everything from bombing missions to the Normandy invasion.

In 1945, as the war effort wound down, scientists wondered what would become of these enterprises. The United States Navy decided to step in and fund basic research through a new Office of Naval Research. The support for science, later taken up by other military services as well, was propelled by a group of officers who saw they would need scientists for many purposes. The war had been shortened, if not decided, by radar, atomic bombs, and dozens of other scientific devices barely imagined a decade earlier. Who could guess what basic research might turn up next? Ready access to skilled brains might be vital in some future emergency. Meanwhile, scientists who made famous discoveries would bring prestige to the nation in the global competition with the Soviet Union that was getting under way—the Cold War. So there was reason to support good scientists regardless of what questions they chose to pursue. Still, some fields of science were more equal than others in the long-term advantages they might provide to the United States.

Physical geoscience was one of the privileged fields. Military officers recognized that they needed to understand almost everything about the environments in which they operated, from the ocean depths to the top of the atmosphere. In view of the complex interconnectedness of all things geophysical, the military services were ready to sponsor many kinds of research. For good practical reasons, then, the U.S. government supported geophysical work in the broadest fashion. If purely scientific discoveries happened along the way, they would be a welcome bonus.

Meteorology was especially favored. The Air Force had a natural concern for the winds and was particularly generous with support, but other military and civilian agencies joined to foster research that might eventually improve weather prediction. Beyond the daily forecast, some experts had visions of deliberately altering the weather. Schemes to provoke rain by "seeding" clouds with silver iodide smoke caught the public's attention in the 1950s, and government officials and politicians took heed. The U.S. government was pressed to fund a variety of meteorological studies in hopes of improving agriculture with timely rains. A nation that understood weather might also obliterate an enemy with droughts or endless snows—a Cold War indeed! A few scientists warned that "climatological warfare" from cloud seeding or the like could become more potent than even nuclear bombs.

These programs all addressed questions of how to predict, and perhaps control, the weather temporarily within a localized region. Questions about long-term climate change over the planet as a whole were *not* a favored field of inquiry. Why pay for research about, for example, the global effects of increased CO_2, when such change was not expected for centuries to come, or more likely never?

Nobody advised Gilbert Plass to study greenhouse warming. The Office of Naval Research supported him to do theoretical calculations for an experimental group at the Johns Hopkins University who were studying infrared radiation. As Plass later recalled, he got curious about climate change only because he read broadly about topics in pure science. He happened upon the discredited theory that the ice ages could be explained by changes in CO_2. Plass took to studying how CO_2 in the atmosphere absorbed infrared radiation, as an adjunct to his official work. Before he finished his analysis, he moved to southern California to join a group at the Lockheed Aircraft Corporation, studying questions of infrared absorption directly related to heat-seeking missiles and other weaponry. Meanwhile he wrote up his results on the greenhouse effect—"in the evening," as he recalled, taking a break from his weapons research.[2]

Plass knew the old objection to the greenhouse theory of climate change: in the parts of the spectrum where infrared absorption took place, the CO_2 and water vapor that were already in the atmosphere sufficed to block all the radiation that could be blocked, so a change in the level of the gas could not matter. Doubts about this had been raised during the 1940s by new measurements and an improved theoretical approach. The old measurements, made at sea-level pressure, in fact told little about the frigid and rarified air in the upper reaches of the atmosphere, where most of the infrared absorption takes place. Up there, the broad bands that entirely blocked radiation at sea level resolved into clusters of narrow spectral lines, like a picket fence with spaces between where radiation could slip through. The new precision measurements at low pressure, backed up by improved theory, suggested that adding more CO_2 might indeed change how much radiation was absorbed.

It was not possible to say anything more specific without extensive computations. Fortunately, Plass had access to the newly invented digital computers. His lengthy calculations demonstrated that adding or subtracting some CO_2 could make a difference, seriously reducing or increasing the amount of radiation that escaped into space from the Earth's surface. In 1956 Plass announced that human activity would raise the average global temperature "at the rate of 1.1 degree C per century."

Plass's computation was too crude to convince other scientists, for he had left out crucial factors such as possible changes in water vapor and clouds. But he did prove a central point: the greenhouse effect could not be dismissed with the old argument that adding more CO_2 could make no difference. He warned that climate change could be "a serious problem to future generations"—although not for several centuries. Like Arrhenius and Callendar, Plass was chiefly interested in the mystery of the ice ages. If the world's temperature continued to rise to the end of the century, he thought it would matter mainly to scientists, by confirming the CO_2 theory of climate change.[3]

Lockheed was not far from the California Institute of Technology, where Dave Keeling was pursuing his own questions about CO_2. He read Plass's work, talked with him, and was impressed. When Keeling had begun studying the fluctuating levels of CO_2 in the atmosphere, he had spoken of possible applications for agriculture. But his true interest was the pure scientific study of geochemistry on a global scale. What processes affected the level of CO_2, and what did that level affect in return?

Answering such questions would need measurements at a level of accuracy beyond what could be reached by any instrument on the market. Keeling spent months of research, ingenuity, and labor building his own instrument. As he measured the air in various locations around California, patiently refining his techniques, Keeling found that at the most pristine locations he kept getting the same number. It must be the true base level of atmospheric CO_2, underlying the passing pulses emitted by an upwind factory or farm. (The Scandinavian scientists who monitored the gas never hoped to see any such thing, and failed to hunt out all the errors in their techniques as Keeling managed to do.)

The next question was whether the level was gradually rising, as Plass and Callendar suspected. That question had little chance to attract research funds. Experts believed any rise of CO_2 would be too slow to matter for centuries to come, and probably couldn't happen at all. For if Plass had shown that the facts of infrared absorption did not rule out greenhouse warming, another weighty objection to the theory remained. Wouldn't the oceans simply swallow up whatever extra CO_2 we humans might pour into the atmosphere?

It happened that the movements of carbon could now be tracked with a new tool, radiocarbon—a radioactive isotope, carbon-14.

Such isotopes had come under intense study during the wartime work to build nuclear weapons, and the pace had not slackened in the postwar years. Sensitive instruments were developed to detect radioactive fallout from Soviet nuclear tests, and a few scientists turned the devices to measuring radiocarbon. Their studies also drew support from interests far removed from the Cold War. Archeologists and the philanthropists who supported them were fascinated by the way radiocarbon measurements could give exact dates for ancient relics such as mummies or cave bones. The isotope is created in the upper atmosphere, when cosmic-ray particles from outer space strike the carbon atoms in carbon dioxide or other gases. Some of the radiocarbon would find its way into living creatures. After a creature's death the isotope would slowly decay away over millennia at a fixed rate. Thus the less of it that remained in an object, in proportion to normal carbon, the older the object was.

As with most of the measurement techniques that play a role in this story, it was a lot easier to describe radiocarbon dating in principle than to carry it out in practice. Any contamination of a sample by outside carbon (even from the researcher's fingerprints) had to be stringently excluded, of course, but that was only the beginning. Delicate operations were needed to extract a microscopic sample and process it. And frustrating uncertainties prevailed until workers understood that their results had to be adjusted for the room's temperature and even the barometric pressure.

One of the new radiocarbon experts, the chemist Hans Suess, thought of applying the technique to the study of geochemistry. It occurred to him that the carbon emitted when humans burned fossil coal and oil is ancient indeed, its radioactivity long gone. He gathered wood from century-old trees and compared it with modern samples. In 1955, Suess announced that he had detected that ancient carbon had been added to the modern atmosphere, presumably from the burning of fossil fuels. But he figured that the added carbon made up barely one percent of all the carbon in the atmosphere—a figure so low that he concluded that most of the carbon derived from fossil fuels was being promptly taken up by the oceans. A decade would pass before he managed to get more accurate measurements, which would show a far higher fraction of fossil carbon.

Everyone knew how tricky it was to measure radiocarbon, and it was obvious that Suess's data were preliminary and insecure. The important thing he had demonstrated was that fossil carbon did show up in the atmosphere. With more work one might figure out exactly how long it took the oceans to absorb the carbon derived from burning fossil fuels. The question was intriguing, for as an oceanographer admitted, "nobody knows whether it takes a hundred years or ten thousand."[4]

This oceanographer was Roger Revelle, a dynamo of a researcher and administrator who was driving the expansion of the Scripps Institution of Oceanography near San Diego, California. Sitting on a dramatic cliff overlooking the Pacific, the prewar Scripps had been a typical oceanographic establishment, quiet and isolated, with its small clique of a dozen or so gossiping researchers and a single research ship. It relied on private patronage, which faltered when the Depression bit into the Scripps family's funds. The postwar Scripps was growing into something quite different, a complex of modern laboratories. Revelle, supplementing the basic support that he got from the public purse through the University of California, won funding for a variety of projects under contracts from the Office of Naval Research, the natural patron for any research related to the oceans, as well as other federal agencies. One of Revelle's many good ideas was to use some of the money to hire Suess to come to Scripps and pursue radiocarbon studies. By December 1955 the two had joined forces, combining their expertise to study carbon in the oceans.

From measurements of radiocarbon in seawater and air, Suess and Revelle deduced that the ocean surface waters took up a typical molecule of CO_2 from the atmosphere within a decade or so. Other scientists who looked into the question around the same time confirmed the conclusion. Yes, the oceans did absorb most of the carbon humanity added to the atmosphere. The only question left, it seemed, was whether it would accumulate near the surface, or whether currents would carry it deep into the oceans.

Revelle's group was already studying the question of how fast the ocean surface waters turned over. It was a matter of national interest, for the Navy and the U.S. Atomic Energy Commission were concerned about the fate of fallout from bomb tests. The Japanese were in an uproar over contamination of the fish they relied upon. Moreover, if the ocean currents were slow enough, radioactive waste from nuclear reactors might be dumped on the seabed. Measurements of radiocarbon at various depths and other studies, pursued at Scripps and elsewhere in the 1950s, showed that on average the ocean waters turn over completely in several hundred years. That seemed fast enough to sweep the CO_2 produced by human industry into the depths.

Revelle liked to pursue various lines of research in parallel, and now another of his interests happened to be relevant. Back in 1946 the Navy had sent Commander Revelle (he had taken the naval rank during the war) to Bikini Atoll, at the head of a team that studied the lagoon in preparation for a nuclear bomb test. The seas are not just salt water but a complex stew of chemicals, and it was not easy to say what adding more chemicals would mean for, say, the carbon in coral. Through the following decade Revelle kept trying out calculations and abandoning them, until one day he realized that the peculiar chemistry of seawater would prevent it from retaining all of the carbon that it might take up.

The mix of chemicals in seawater creates a buffering mechanism that stabilizes the water's acidity. This had been known for decades, but nobody had realized what it meant for CO_2. Now Revelle saw that when some molecules were absorbed, their presence would alter the balance through a chain of chemical reactions. While it was true that most of the CO_2 molecules added to the atmosphere would wind up in ocean surface water within a few years, most of these molecules (or others already in the oceans) would promptly be evaporated. Revelle calculated that in sum, the ocean surface could not really absorb much gas—barely one-tenth the amount earlier calculations had predicted. Whatever CO_2 humanity added to the atmosphere would not be swallowed up promptly, but only over thousands of years.

It was now 1957, and Revelle's radiocarbon paper with Suess was all written and ready to send off for publication. Revelle went back and made a few changes. Most of the text he left as it was, reflecting the pair's original belief that the oceans were absorbing most of the new CO_2. A few extra sentences explained that this would not happen after all. As sometimes happens with landmark scientific papers, written in haste while understanding just begins to dawn, Revelle's explanation was so obscure that it took other scientists a couple of years to understand and accept it. Revelle himself did not quite grasp all the implications. Along with the brief explanation, he published a quick calculation which indicated that the amount of CO_2 in the atmosphere would gradually rise over the next few centuries, then level off with a total increase of 40 percent or less.

That reassuring conclusion was a gross underestimate. Revelle was assuming that industrial emissions during the coming centuries would continue at the 1957 rate. Scarcely anyone had yet grasped the prodigious fact that both population and industrialization were exploding in exponential growth. Between the start of the twentieth century and its end the world's population would quadruple, and the use of energy by an average person would quadruple, making a 16-fold increase in the rate of emission of CO_2. Yet at mid-century, world wars and the Great Depression had led most technologically advanced nations to worry about a possible *decline* in their populations. Their industries seemed to be plodding ahead, expanding no faster in the current decade than in the last decade. As for "backward" regions like China or Brazil, industrialization had scarcely entered anyone's calculations except as a possibility for the remote future.

Different ideas were beginning to stir. In particular, Keeling's mentor at CalTech, the geochemist Harrison Brown, was sketching out a more realistic vision of a future with exploding population and industrialization. Revelle had heard these ideas, and before he sent off for publication the paper he had written with Suess, he added a remark: the accumulation of CO_2 "may become significant during future decades if industrial fuel combustion continues to rise exponentially." In conclusion, he wrote, "Human beings are now carrying out a large scale geophysical experiment of a kind that could not have happened in the past nor be reproduced in the future."[5]

Revelle meant "experiment" in the traditional scientific sense, a nice opportunity for the study of geophysical processes. Yet he did recognize that there might be some future risk. Other scientists too began to feel a mild concern as they gradually assimilated the meaning of Plass's and Revelle's difficult calculations. Adding CO_2 to the atmosphere could change the climate after all. And the changes might arrive not in some remote science-fiction future, but within the next century or so.

Revelle and Suess, like Arrhenius, Callendar, Plass, and everyone else who up till then had made a contribution toward the discovery of global warming, had taken up the question as a side issue. They saw in it a chance for a few publications, a detour from their main professional work, to which they soon returned. If just one of these men had been possessed by a little less curiosity, or a little less dedication to laborious thinking and calculation, decades more might have passed before the possibility of global warming was noticed. It was also a historical accident that military agencies were scattering money with a free hand in the 1950s. Without the Cold War there would have been little funding for the research that turned out to illuminate the CO_2 greenhouse effect, a subject nobody had connected with practical affairs. The U.S. Navy had bought an answer to a question it had never thought to ask.

Revelle and Suess were now eager to learn more about the "large scale geophysical experiment" of greenhouse warming. Few others paid much attention to their paper, with its complex and difficult technical argument. Government agencies like the Navy could hardly be expected to devote much more money to the question. Not only did it seem remote from anything they needed to know, but it was unlikely to yield scientific information without great effort. Fortunately, just then another purse opened up. The new funds came (or seemed to come) from peaceful internationalism, altogether apart from national military drives.

Geophysics is inescapably international. Ocean currents and winds flow each day between regions. Yet up through the mid-twentieth century most geophysicists studied phenomena within a region, often not even a nation but part of a nation. Meteorologists of different nationalities did cooperate in the loose informal fashion of all science, reading one another's publications and visiting one another's universities. Gradually, their subject drove them to join forces more closely than did most scientists. In the second half of the nineteenth century, leaders of the field got together in a series of international congresses which led to the creation of an International Meteorological Organization. Similar organizational drives were strengthening the entire area of geophysics, including most of the fields relevant to the study of climates. Already in 1919 an International Union of Geodesy and Geophysics had been founded, with separate sections for different fields such as oceanography. An American Geophysical Union was also created in 1919, followed by other national societies and a few journals such as the German *Zeitschrift für Geophysik*. But these early institutions were too weak to provide strong connections. Most individuals who might be called geophysicists did the bulk of their work within the confines of one or another single field such as geology or meteorology. And nearly every research project was carried out entirely within one or another particular nation.

After the Second World War, governments saw new reasons to support international cooperation in science. This was the era that saw the creation of the United Nations, the Bretton-Woods financial institutions, and many other multilateral efforts. The aim was to bind peoples together with interests that transcended the self-serving nationalism that had brought so much horror and death. When the Cold War began it only strengthened the movement, for while tens of millions had recently been slaughtered, nuclear arms could slay hundreds of millions. It seemed essential to create areas where cooperation could flourish. Science, with its long tradition of internationalism, offered some of the best opportunities. Fostering transnational scientific links became an explicit policy for the world's leading democracies, not least the United States. It was not just that gathering knowledge gave a handy excuse for creating international organizations. Beyond that, the ideals and methods of scientists, their open communication, their reliance on objective facts and consensus rather than command, would reinforce the ideals and methods of democracy. As the political scientist Clark Miller explained, American foreign policy makers believed the scientific enterprise was "intertwined with the pursuit of a free, stable, and prosperous world order."[6]

Study of the global atmosphere seemed a natural place to start. In 1947 a World Meteorological Convention explicitly made the meteorological enterprise an intergovernmental affair. In 1951 the International Meteorological Organization was succeeded by the World Meteorological Organization, or WMO (one of the first of the acronyms that would infest everything geophysical). The WMO, an association of national weather services, soon became an agency of the United Nations. That gave meteorological groups access to important organizational and financial support, and lent them a new authority and stature. But all this did little to connect together the scattering of scientists in diverse fields who took an interest in climate change.

There had never been a community of scientists studying climate change. Studies were pursued only by individuals with one or another interest who turned their attention, for a few years, to some special aspect of the topic. An astrophysicist studying changes in solar energy, a geochemist studying the movements of radioactive carbon, and a meteorologist studying the global circulation of winds had little knowledge and expertise in common. Even within each field, specialization often separated people who might have had something to teach one another. They were unlikely to meet at a scientific conference, read the same journals, or even know of one another's existence. Revelle's decision to bring Suess to Scripps and join forces, oceanographer with geochemist, was a well-conceived exception to the rule.

By the mid-twentieth century, few scientists managed to carry out significant work in more than one field. The knowledge you needed had become too deep, and the techniques too esoteric. Trying to become expert in a second field of knowledge diverted energy and risked your career. "Entering a new field with a degree in another is not unlike Lewis and Clark walking into the camp of the [Native American] Mandans," remarked Jack Eddy, a solar physicist who took up the study of climate change. "You are not one of them . . . Your degree means nothing and your name is not recognized. You have to learn it all from scratch."[7]

To make communication still harder, different fields attracted different kinds of people. If you went into the office of a statistical climatologist, you could expect to find ranges of well-organized shelves and drawers stacked with papers bearing neat columns of figures. In later years the stacks would hold computer printouts, the fruit of countless hours spent coding programs. The climatologist was probably the kind of person who, as a boy, had set up his own home weather station and meticulously recorded daily wind speed and rainfall, year after year. Go into the office of an oceanographer and you were more likely to find a jumble of curiosities from the shores of the seven seas. You could hear adventure stories, like one experienced scientist's tale of how he was washed overboard and escaped drowning by a hair. Oceanographers tended to be salty types, accustomed to long voyages far from the comforts of home, outspoken and sometimes self-centered.

These differences went along with divergence in matters as fundamental as the sorts of data people used. Climate experts, for example, relied upon the WMO's world-spanning network of thousands of weather stations where technicians reported standardized data. Oceanographers personally built their instruments and lowered them over the side of one of their few research ships. The climatologist's weather, constructed from a million numbers, was something entirely different from the oceanographer's weather—a horizontal blast of sleet or a warm relentless trade wind. There were even technical divergences. As one climate expert remarked in 1961, "The fact that there are so many disciplines involved, as for instance meteorology, oceanography, geography, hydrology, geology and glaciology, plant ecology and vegetation history—to mention only some—has made it impossible to work . . . with common and well established definitions and methods."[8]

Such fragmentation was becoming intolerable. In the mid-1950s a small band of scientists worked out a scheme to boost cooperation among the various geophysics disciplines. They hoped to coordinate data-gathering on an international scale and—no less important—to persuade governments to add an extra billion or so dollars to their funding of geophysics research. They succeeded with the creation of the International Geophysical Year of 1957–58. The IGY would draw together scientists from many nations and a dozen different disciplines, to interact in committees that would plan and carry out interdisciplinary research projects grander than any attempted before.

A variety of motives converged to make the IGY possible. The scientist organizers hoped chiefly to advance scientific knowledge, and thereby to advance their individual careers. The government officials who supplied the money, while not indifferent to pure scientific discovery, expected the new knowledge would have civilian and military applications. The American and Soviet governments and their allies further hoped to win practical advantages in their Cold War competition. Under the banner of the IGY they could collect global geophysical data of potential military value. Along the way each nation hoped to gather intelligence about its rivals, and meanwhile enhance its own prestige. Some scientists and officials, conversely, hoped that the IGY would help set a pattern of cooperation between the opposed powers—as indeed it would. It is a moot question whether, in a more tranquil world, governments would have spent so much to learn about seawater and air. For whatever motives, the result was a coordinated effort involving several thousand scientists from 67 nations.

Climate change ranked low on the list of IGY priorities. But with such a big sum of new money there was bound to be something for climate-related topics. The study of CO_2 was one minor example. In the committees that allocated the U.S. share of funding, Revelle and Suess argued for a modest program to measure the gas in the ocean and air simultaneously at various points around the globe. It wouldn't cost much, so the committee granted some money. Revelle already had Keeling in mind for this work, and he now hired the young geochemist to come to Scripps and conduct the world survey. Revelle aimed to establish a baseline "snapshot" of CO_2 values around the world, averaging over the large variations observed from place to place and from time to time. After a couple of decades, somebody could come back, take another snapshot, and see if the average CO_2 level had risen.

Keeling aimed to do better than that. "Keeling's a peculiar guy," Revelle later remarked. "He wants to measure CO_2 in his belly . . . And he wants to measure it with the greatest precision and the greatest accuracy he possibly can."[9] The greatest accuracy called for expensive new instruments, far more precise than most experts thought were called for to measure something that fluctuated as widely as CO_2 levels. Keeling lobbied key officials and managed to persuade them to give him money for the instruments. He set one up atop the volcanic peak Mauna Loa in Hawaii, surrounded by thousands of miles of clean ocean, one of the best sites on Earth to measure the undisturbed atmosphere. Another instrument went to the even more pristine Antarctic (where researchers depended wholly on military logistical support, in the literally coldest site of Cold War activity).

Keeling's costly equipment, together with his relentless pursuit of every possible source of error, paid off. In Antarctica, he tracked down variations in the CO_2 measurements to emissions from nearby machinery. On Mauna Loa, gas leaking from vents in the volcano itself was to blame. Stalking such problems with meticulous attention to detail, Keeling nailed down a remarkably precise and consistent baseline number for the level of CO_2 in the atmosphere. His first twelve months of Antarctic data hinted that a rise could be seen in just that one year.

But the IGY was winding down. By November 1958, the remaining funds had fallen so low that CO_2 monitoring would have to stop. Keeling scrambled to find more money. Suess and Revelle diverted a fraction from a grant that the Atomic Energy Commission had given Scripps for other purposes (in those days, more than now, agencies trusted scientists to spend funds as they chose). In 1960, with two full years of Antarctic data in hand, Keeling reported that the baseline CO_2 level had risen.[10] The rate of the rise was approximately what would be expected if the oceans were not swallowing up most industrial emissions.

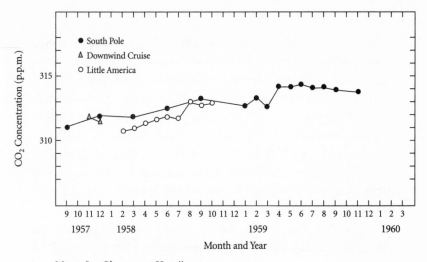

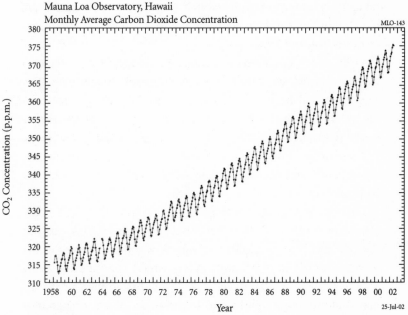

Figure 1. THE RISE OF CARBON DIOXIDE IN THE ATMOSPHERE.

Upper: A rising level of CO_2 in the atmosphere was first demonstrated in 1960 in Antarctica, visible after only two years of measurements. (C. D. Keeling, *Tellus* 12, p. 200, 1960, reproduced by permission.) *Lower:* The "Keeling curve" of CO_2 measured at Mauna Loa, Hawaii, over nearly half a century. Within the long-term rise are annual fluctuations as Northern Hemisphere plants take up carbon during summer growth and release it in winter decay. Note the hiatus when funds ran out in spring 1964. (Scripps Institution of Oceanography, reproduced by permission.)

In 1963 the funds ran out altogether, and CO_2 monitoring shut down. Although some scientists had immediately recognized the importance of Keeling's work, no agency felt responsible for funding a climate study that might run for many years. Meanwhile Keeling had applied to the National Science Foundation, a U.S. federal agency established back in 1950 on a modest budget. The NSF's situation had improved in 1958 after the launching of the Soviet Sputnik and other satellites. To the American public, Sputnik was a frightening demonstration of vulnerability to nuclear-armed missiles, and seemed to show a Russian lead in science and technology. The government quickly boosted funding for all areas of science. The NSF, its wallet full, took over from military agencies much of the nation's support of basic research. One minor consequence was that Keeling got funds to continue the Mauna Loa measurements after only a short hiatus.

As the Mauna Loa data accumulated, the record grew increasingly impressive, showing CO_2 levels noticeably higher year after year. What had begun as a temporary job for Keeling was turning into a lifetime career—the first of many careers that scientists would eventually dedicate to climate change. Within a few years, Keeling's inexorably rising CO_2 curve was widely cited by scientific review panels and science journalists. It became the central icon of the greenhouse effect.

Keeling's data put the capstone on the structure built by Tyndall, Arrhenius, Callendar, Plass, and Revelle and Suess. This was not quite the discovery of global warming. It was the discovery of the *possibility* of global warming. Experts would continue for many years to argue over what would actually happen to the planet's climate. But no longer could a well-informed scientist dismiss out of hand the possibility that our emission of greenhouse gases would warm the Earth. That odd and unlikely theory now emerged from its cocoon, taking flight as a serious research topic.

CHAPTER THREE

A DELICATE SYSTEM

The weather means a lot to people in Boulder, Colorado, a city of mountaineers and skiers, including not a few scientists. The winter winds there can nudge cars off the road, and in summer you can sit in the hills above town to watch thunderstorms sail across the high plains, churning with lightning. One of the most striking sights in Boulder is the sandstone-red towers of the National Center for Atmospheric Research. Created by Congress in 1960 under pressure from scientists seeking a way around the Weather Bureau's cobwebs, NCAR was dedicated to the scientific study of everything in the atmosphere from cloudbursts to climate. Boulder was a good spot for the conference on "Causes of Climate Change" that convened there in August 1965. The meeting was scarcely noticed by most scientists at the time, but in retrospect it was a turning point.

The organizers had deliberately brought together experts in everything from volcanoes to sunspots, presided over by the oceanographer Roger Revelle. Lectures and roundtable discussions were full of spirited debate as rival theories clashed, and Revelle needed all his exceptional leadership skills to keep the meeting on track. The conference was convened mainly to discuss the many rival explanations of the ice ages in the comfortable traditional mode. Instead, it exploded with new ideas that pointed to a novel and foreboding way of looking at the future of climate. The planet's climate, the scientists agreed, could not be treated in the old fashion like some simple mechanism that kept itself stable. It was a complex system, precariously balanced. The system showed a dangerous potential for dramatic change, on its own or under human technological intervention, and quicker than anyone had supposed.

At the Boulder meeting not only the climate, but ways of studying it, appeared in a new light. The familiar unchanging climatology of statistical compilations held no appeal for these scientists. They were trying to build up their knowledge from solid mathematics and physics, aided by new techniques drawing on fields from microbiology to nuclear chemistry. But science alone could not explain the deep shift in views about one of the fundamental components of human experience. Events had been altering the thinking of everyone in modern society.

Is human technology a force of geophysical scope, capable of affecting the entire globe? Surely it is not, thought most people in 1940. Surely it is, thought most in 1965. The reversal was not because of any changes in what scientists knew about global warming. The public's rising concern for human impacts came from more visible connections between technology and the atmosphere. One of these was a growing awareness of the dangers of atmospheric pollution. In the 1930s, citizens had been happy to see smoke rising from factories: dirty skies meant jobs. But in the 1950s, as the economy soared and life expectancy lengthened in industrialized countries, a historic shift began, from worries about poverty to worries about chronic health conditions. Doctors were learning that air pollution was mortally dangerous for some people. Meanwhile, in addition to smoke from coal-burning factories came exhaust from the rapidly proliferating automobiles. A "killer smog" that smothered London in 1953 demonstrated that the stuff we put into the air could actually slay several thousand people in a few days.

The public's attention was also drawn to the air by the news of attempts to make rain by "seeding" clouds. Scientists openly speculated about other technical tricks, such as spreading a cloud of particles at a selected level in the atmosphere to interfere with solar radiation. Journalists and science-fiction writers suggested that with such techniques, the Russians might someday inflict deadly blizzards on the United States. It had become plausible that by putting materials into the air humans could alter climate on the largest scale, perhaps not for the better.

The biggest stimulus to changes in thinking was the astonishing advent of nuclear energy. Suddenly nothing seemed beyond human power. To many people the news of a limitless energy source was hopeful, even utopian. Among many other wonders, experts speculated about salvoes of atomic bombs to control weather patterns, bringing rain exactly where it was needed. At the same time, scientists warned that a nuclear war could destroy civilization. Widely seen movies and novels pictured the extinction of all life by radioactive fallout, carried around the world on the winds after a nuclear war.

By the late 1950s utopian hopes about technology began to dissolve as the nuclear arms race accelerated. Rising fears found a voice in shrill public debates and mass demonstrations against nuclear weapons tests. Exquisitely sensitive instruments could detect radioactive fallout from test explosions half a world away—the first recognized form of global atmospheric pollution. Then in 1962 Rachel Carson published *Silent Spring,* warning that pesticides such as DDT and other chemical pollution, drifting around the world much like fallout, could endanger living creatures not just in the neighborhood of the polluter, but everywhere. Feelings of dread multiplied: whether or not technology would turn deserts into gardens, it could demonstrably turn gardens into deserts!

Many among the public suspected that dust from atomic bomb tests was already affecting the weather. From about 1953 until open-air testing ceased in the mid-1960s, as opponents of nuclear armaments pointed with horror to the invisible dangers of fallout, some people blamed the faraway tests for almost any unseasonable heat or cold, drought or flood. In a magazine article laying out the evidence that global temperatures were rising, the authors remarked that "Large numbers of people wonder whether the atomic bomb is responsible for it all."[1]

The new threats awoke images and feelings that most people had scarcely experienced outside their dreams and nightmares. Humans were introducing unnatural technologies, meddling with the very winds and rain, spreading pollution everywhere. Would we provoke retribution? Would "Mother Nature" pay us back for our attacks upon her? Such veiled anxieties were not detectable in the sober discussions of subjects like climate change. Outside of the nuclear and chemical controversies, hardly anyone used the charged language of defilement and transgression. But the public did develop a vague feeling that natural disasters followed not only scientific law but moral law—a punishment for unhallowed human assaults.

Of course, this was nothing new. Many tribal peoples attributed climate disasters, such as an unusually bad winter, to human transgression. Somebody's "polluting" violation of a taboo was to blame. But during the 1950s, tales of various sorts of affliction all the way up to world catastrophe took on a veneer of scientific plausibility. As the nuclear arsenals grew, biblical fundamentalists got a wider hearing than ever for their prophecies of rains of fire, rivers of blood, and so on, with wars and sin ushering in the Apocalypse.

Now that it seemed plausible that human technology could alter the planet as a whole, journalists found it easier to suggest that burning fossil fuels could change the climate. Evidence that the world had been growing measurably warmer had become strong enough to convince most meteorologists. During the 1950s, newspaper readers repeatedly saw small items reporting anecdotes of warming, especially in the Arctic. For example, in 1959 the *New York Times* reported that the ice in the Arctic Ocean was only half as thick as it had been in the previous century. Still, the report concluded, "the warming trend is not considered either alarming or steep."[2] Compared with chemical pollution and nuclear war, climate change was an old-fashioned issue, and most likely harmless.

Revelle took the lead in suggesting that trouble might lie ahead. As soon as he calculated that a rise in the CO_2 level was likely, Revelle took pains to talk about global warming with science journalists and government officials. Noting that climate had changed abruptly in the past, perhaps bringing the downfall of entire civilizations in the ancient world, he warned that the CO_2 greenhouse effect might turn Southern California and Texas into "real deserts." Testifying to Congress in 1956 and 1957, he was one of the first to use a new and potent metaphor: "The Earth itself is a spaceship," he said. We had better keep an eye on its air control system.[3]

Revelle did not expect much change in the climate for many decades, and perhaps never. He only meant to prod the government to pay for the IGY and other geophysical research in general. Few others foresaw anything ominous at all. "There would seem to be every reason for producing as much carbon dioxide as we can manage," one popularization concluded. "It is helping us towards a warmer and drier world."[4] In any case nothing would happen until the twenty-first century—which seemed very distant indeed from the 1950s. The subject was scarcely noticed by anyone outside the science-minded minority who happened upon reports by science journalists. Those were mostly buried in the back pages of newspapers or dropped into a news magazine as a brief paragraph or two.

The one fact that stubbornly could not be overlooked by those who paid any attention was the climbing level of CO_2 in the atmosphere. Keeling's curve rose year by year through the 1960s. A few scientists began thinking that somebody should actually do something about the matter—for a start, take up climate research more systematically.

The first step in that direction came in 1963, when Keeling and a few other experts met in a conference sponsored by the private

Conservation Foundation. They issued a report suggesting that the doubling of CO_2 projected for the next century could raise the world's temperature by 4°C (more than 6°F). They warned that this could be harmful; for example, it could cause glaciers to melt and raise the sea level so that coastlines would get flooded. The federal government should give the subject more consistent attention, they said, with better organization and more money.[5]

The government reacted to such complaints at its usual deliberate pace. In 1965, when the President's Science Advisory Committee formed a panel to address environmental issues, it included a subpanel of climate experts. They reported that greenhouse-effect warming was a matter of real concern. That put the issue on the official agenda at the highest level—although only as one item on a long list of environmental problems, many of which seemed more pressing. The next step in such matters was typically to ask the National Academy of Sciences to form a committee and issue an authoritative report. In 1966 the Academy duly pronounced on how human activity could influence climate. The experts sedately said there was no cause for dire warnings, but they did believe the CO_2 buildup should be watched closely. "We are just now beginning to realize that the atmosphere is not a dump of unlimited capacity," the report said, "but we do not yet know what the atmosphere's capacity is."[6] The primary conclusion was typical of such reports—a maxim that came from the heart of scientists' belief in their calling—More Money Should Be Spent on Research.

That was easier said than done, for research on climate change was not on any official's list of responsibilities. As the 1965 advisory panel remarked, "no agency or program is concerned with the average condition of our environment."[7] The 1966 Academy report added that for climate as for most environmental fields, "there exists no single natural advocate in the Federal structure, nor is there a clear mechanism for making budgetary decisions." In the mid-1960s a variety of government agencies together spent roughly $50 million a year for all aspects of meteorological research. That was not much, and climate change caught only a small percentage of that sum.[8] Other nations were spending even less. Studies of climate had to fit in as minor components of programs that had been set up to work on other problems.

Priority was given to predicting weather a few days ahead. When thoughts turned to longer terms, the titles of the panels show what was on people's minds. The President's advisory group was named the "Environmental Pollution Panel," and the Academy's was the "Panel on Weather and Climate Modification." Asked about human influence on the atmosphere, the public would think first about smog, and next about deliberate attempts to make rain. Such attempts included climatological warfare—indeed the U.S. armed forces had already secretly begun a large-scale attempt to bog down the North Vietnamese army with artificial rainmaking. Progress on global warming would come mainly from money earmarked for other purposes.

In climate studies the great prize, still unclaimed after a century of work, was the explanation of the ice ages. The best hopes for progress lay in new techniques. The most precise and ingenious one had been invented early in the century by a few Swedish scientists: the study of ancient pollens. The tiny but amazingly durable pollen grains are as various as sea shells, with baroque lumps and apertures characteristic of the type of plant that produced them. One could dig up soil from lake beds or peat deposits, dissolve away in acids everything but the sturdy pollen, and after some hours at a microscope know what kinds of flowers, grasses, or trees had lived in the neighborhood at the time the layer was formed. That told scientists much about the ancient climate: we had no readings from rain gauges and thermometers 50,000 years ago, but pollen served as an accurate proxy. Pollen data were invaluable for identifying strata as an aid to oil exploration, and that paid for specialists to bring the technique to a high degree of refinement.

The usefulness of pollen and other proxies for ancient climate was redoubled by the new technique of radiocarbon dating. Go to a glacial moraine or lake bed, dig out fragments of ancient trees, measure the fraction of radioactive carbon in them, and you had a date. Thus a reliable timescale could now be assigned to the climate fluctuations that formerly had been only sketched out by the proxies. For example, dating of lake deposits in the western United States showed highly regular cycles of drought and flood. It was difficult to match these with the geologists' traditional sequence of ice-age cold and warm periods. Oddly, they seemed to match the 21,000-year cycle predicted by Milankovitch's astronomical calculations of shifts in sunlight.

Another promising new technique likewise came from a surprising combination of biology and nuclear science. The oceans swarm with microscopic plankton, including countless foraminifera (nicknamed forams), single-celled animals that scavenge with pseudopods poking out through holes in their shells. When forams die, their tiny shells drift down into the ooze of the seabed and there endure for ages, in some places so numerous that they form thick deposits of chalk or limestone. In 1947 the nuclear chemist Harold Urey found a way to take the temperature of an ancient ocean by measuring the oxygen that forams built into their shells. The rare isotope oxygen-18 is a bit heavier than normal oxygen-16, and biologists had demonstrated that the amount of each isotope that a foram takes up varies with the temperature of the water. The isotopes were fossilized with the shells, and the ratio of isotopes (O^{18}/O^{16}) could be measured with the new and highly sensitive tools developed for nuclear studies.

The method was taken up by Cesare Emiliani, a geology student from Italy working in Urey's laboratory at the University of Chicago. There were many problems to solve before he could get reliable results. (Urey, already a Nobel Prize winner, called it "the toughest chemical problem I ever faced.")[9] And before the chemistry could even begin, samples had to be extracted from the sticky sediments of the ocean floor without disturbing the layers. Börge Kullenberg—let us remember here the name of at least one of the many people who contributed a modest but essential technique to this history—solved the problem for a Swedish Deep Sea Expedition in 1947. He put a piston inside a long tube and pulled the piston up to suck in the sediment while the tube was being shoved into the seabed. Kullenberg could recover cylindrical cores more than 20 meters long. Back in the lab, somebody would put a sample of the muck under a microscope and tease out a few hundred of the shells, each no bigger than the period at the end of this sentence. The shells were ground to powder, which was roasted to extract CO_2 gas so the isotopes could be measured. Technicians needed to be very careful to avoid contamination by any other source of gas, such as their own breath.

In 1955 Emiliani applied these exacting methods to slimy cylinders of mud and clay totaling hundreds of meters in length, extracted from the deep seabed by various expeditions in recent years and carefully stored away in oceanographic institutions. Borrowing cores from several expeditions, he assembled a remarkable record of temperature changes stretching back nearly 300,000 years. To get a timescale connecting the temperature changes with depth down the core, he made radiocarbon measurements covering the top few tens of thousands of years (farther back there was too little of the isotope left to measure). That gave him an estimate for how fast sediments accumulated on the seabed at that point. He found a rough correlation with Milankovitch's astronomical calculations. Temperatures seemed to rise and fall in time with the varying amount of sunlight that struck high northern latitudes in summer.

If his curves did fit Milankovitch's, then they could not well match the sequence of ice ages that had been painstakingly worked out by nineteenth-century geologists and accepted ever since. To resolve the issue, Emiliani began urging colleagues to launch a major program and pull up truly long cores, a single hundred-meter record covering many hundreds of thousands of years. But for a long time the drillers' crude techniques were unable to extract long, undisturbed cores from the ooze. As one of them remarked ruefully, "one does not make wood carvings with a butcher's knife."[10] Meanwhile Emiliani, extracting data from the best available cores, announced that he could not fit the traditional ice-ages timetable at all. He rejected the entire scheme, enshrined in textbooks, of four major glacial advances alternating with long and equable interglacial periods. Emiliani's data told a tale of dozens of briefer advances and interglacials, complicated by irregular rises and falls of temperature along the way. His data correlated rather well with the complex Milankovitch curve for summer sunlight at high northern latitude.

Geologists defended their traditional chronology passionately and skillfully. For a while they held their ground. It turned out that Emiliani's data on oxygen isotopes in foram shells did not directly measure ocean temperatures after all. Instead, as other workers in the late 1960s demonstrated, the isotope ratio changed mainly for another reason. When water was withdrawn from the oceans to form continental ice sheets, the isotopes evaporated and then fell as rain or snow in different proportions. In particular, a heavier isotope was less likely to evaporate from the ocean surface than a lighter one. When snowfall built up continental ice sheets, the process had withdrawn from the oceans more of the lighter isotope than the heavier one. Thus no matter what the temperature of the water where the forams lived, during a glacial period their shells wound up with less of the lighter isotope. The changes that Emiliani had detected reflected mainly the changing volume of the planet's ice sheets.

Emiliani defended his results fiercely, loath to admit error. By the end of a decade of debate, all his colleagues rejected his temperatures. But they were quick to acknowledge that his work, error and all, was still a landmark. If it was the growth and decay of ice sheets that changed the ratio of isotopes, the rises and falls of Emiliani's curves still did show the rhythm of the ice ages. There was nothing extraordinary in such a combination of discovery and error. Every great scientific paper is written at the outside edge of what can be known, and deserves to be remembered if there is a nugget of value amid the inevitable confusion.

Clear confirmation of Emiliani's basic discovery came when scientists peered through microscopes to take a census of the particular species of foraminifera, layer by layer down the cores. The assemblage of species varied with the temperature of the water where they had lived. The changes matched Emiliani's curves: the temperatures had indeed changed. Evidently he was right to claim that during the past couple of million years there had been dozens of major glaciations, roughly agreeing with Milankovitch's schedule.

Yet how reliable was any conclusion read from the seabed ooze? As the debate over oxygen isotopes showed, what looks like a simple measured fact can be misleading. Scientists therefore rarely accept a result until it is confirmed by wholly different means. A new voice brought this confirmation, a voice that over the coming decades would increasingly command attention. It belonged to Wallace (Wally) Broecker of the Lamont Geological Observatory. Lamont scientists, isolated amid woods overlooking the Hudson River, were combining geological interests with oceanography and the new radioactive and geochemical techniques in a burst of creative research.

In the late 1960s Broecker and a few colleagues traveled to the tropics to hike around ancient coral reefs. Perched at various elevations above the present sea level, the reefs showed how the oceans had risen and fallen as ice sheets built up on the continents and melted away. The reefs could be dated by hacking out samples and measuring their uranium and other radioactive isotopes. These isotopes decayed over millennia on a timescale that had been accurately measured in nuclear laboratories, and unlike radiocarbon, the decay was so slow that there was still enough left to measure after hundreds of thousands of years. Again Milankovitch's orbital cycles emerged, plainer than ever.

Scientists did not immediately convert to the astronomical-orbit theory. Scientists rarely label a proposed answer to a scientific question "true" or "false," but rather consider how likely it is to be true. Normally a new body of data will shift opinion only in part, making the idea seem a bit more likely or less likely. At the conference on "Causes of Climate Change" held in Boulder in 1965, Broecker would only claim that "The Milankovitch hypothesis can no longer be considered just an interesting curiosity."[11] A few years later, after he and his collaborators accumulated more evidence, they still would not claim to have proved anything for certain. "The often-discredited hypothesis of Milankovitch," they said in 1968, "must be recognized as the number-one contender in the climatic sweepstakes."[12] Some disagreed, reserving the top spot for their own favorite hypotheses.

No matter how strong the match between ancient climate data and orbital cycles, scientists would not find the astronomical theory of climate change really plausible until they had an explanation for how it could physically work. Nobody had forgotten the fundamental objection: the changes in sunlight as the Earth's orbit shifted were trivial. How could a minor change in the forces that bore on the atmosphere make entire continental ice sheets wax and wane? Besides, why should an increase in sunlight at high latitudes in the Northern Hemisphere bring on a worldwide change, melting the ice in the Southern Hemisphere at the same time? If scientists gave serious consideration to Broecker's announcements, that was because other developments were already pushing them to rethink the climate system's fundamental nature.

There had always been meteorologists who challenged common

ways of thinking, and some of their speculations had stubbornly refused to go away. For example, back in 1925 a respected climate expert, C. E. P. Brooks, had proposed that ice ages could begin or end
almost arbitrarily. He began with the familiar idea that an increase
of snow cover would reflect more sunlight, which would further
cool the air. Frigid winds, Brooks suggested, would flow into ad-
joining regions, and the snows would swiftly advance to lower latitudes. Only two stable states of the polar climate were possible, he
asserted—one with little ice, the other with a vast white cap on the
planet. A shift from one state to the other might be caused by some
comparatively slight perturbation. There could be an abrupt and
catastrophic rise or fall of tens of degrees, "perhaps in the course of
a single season."[13]

Most experts scorned such talk, which seemed only too likely
to reinforce notions popularized by religious fundamentalists who
were in open conflict with the scientific community. Believers in
the literal truth of the Bible insisted that the Earth was only a few
thousand years old, and defended their faith by claiming that ice
sheets could form and disintegrate in mere decades. Meanwhile the
widely read pseudo-scientific theories of Immanuel Velikovsky and
others added to public belief in apocalyptic climate upheavals.

Scientists replied by pointing to the detailed climate records that
were recovered from layers of silt and clay. Analysis showed no
changes in less than several thousand years. The scientists failed to
notice that most cores drilled from the seabed could not in fact record a rapid change. For the ooze is constantly stirred by burrowing
worms, as well as by currents and slumping, which blur any abrupt
differences between layers.

Ancient lakes and bogs retained a more detailed record, which
scientists could read in the embedded pollen. Major changes in the
mix of plants that showed up in a given lake suggested that the
last ice age had not ended with a uniformly steady warming, but
with some peculiar oscillations of temperature. A particularly strik-

ing event, identified in the 1930s in Scandinavia, was a warm period followed by a prolonged spell of bitterly cold weather. The cold spell was dubbed the Younger Dryas after *Dryas octopetala,* a delicate but hardy little Arctic flower whose pollen gave witness to frigid tundra. After that period a more gradual warming followed. In 1955 the timing was pinned down by a radiocarbon study which revealed that the chief oscillation of temperatures lay around 12,000 years ago. The change had been rapid—where "rapid," for climate scientists at the time, meant something that took as little as one or two thousand years to change. Most scientists thought the event, if real at all, was just a local Scandinavian happenstance.

In 1956 Hans Suess applied his radiocarbon expertise to determine the dates of fossil shells in deep-sea clay drilled up by Lamont Observatory oceanographers. He reported that the last glacial period had ended with a "relatively rapid" rise of temperature—about 1°C per thousand years. Lamont scientists took their own look at the foraminifera shells, sorting them according to whether the species lived in warm or cold waters, and reported an even more abrupt rise. About 11,000 years ago, they said, the climate had shifted from fully glacial conditions to modern warmth within as little as a thousand years. They acknowledged this was "opposed to the usual view of a gradual change."[14]

Emiliani, always prepared to spar with Lamont scientists, disagreed. He published reasons to believe that the temperature rise of some 8°C had been the expected gradual kind, stretching over some 8,000 years. After considerable public debate, he won his point: the variation in the Lamont group's data did not in fact represent a temperature shift. Like some other sudden changes reported in natural records, it reflected peculiarities in the method of analyzing samples, not the real world itself.

Yet in science even mistakes can be valuable if they set scientists to thinking about overlooked possibilities. The apparent evidence that climate had changed drastically in as little as a thousand years provoked a few people to try to figure out what could have happened. Broecker was at this time still a lowly graduate student at Lamont. He put a bold idea into his doctoral thesis. Looking at the studies that would later be found mistaken plus a variety of other data, he saw a pattern of glacial oscillations that was very different from the accepted gradual wave-like rise and fall. Almost in Brooks's words, Broecker suggested that "two stable states exist, the glacial state and the interglacial state, and that the system changes quite rapidly from one to the other."[15] It was just one comment in a thick and little-read doctoral thesis, and it sounded like Brooks's discredited speculations. Most scientists agreed with Emiliani that the fluctuating data, if accurate at all, reflected local oddities. After all, ice sheets kilometers thick would require thousands and thousands of years to build up or melt away. The physics of ice, at least, was simple and undeniable.

A few were not so sure, in particular Broecker's boss—Lamont's founding director, the brash and autocratic Maurice Ewing. Was there any mechanism, he wondered, that could cause climate to swing rapidly between warm and glacial states? Ewing and a colleague, William Donn, had some ideas about how that might actually happen.

Ewing and Donn were probably stimulated by some strange ideas that Russian scientists were talking about. The starting point was a decades-old legacy of grandiose Soviet engineering dreams. Why not divert Siberian rivers, taking water that flowed uselessly into the Arctic Ocean and sending it south to turn the parched soils of central Asia into farmlands? A few scientists pointed out that the diversion of fresh water would make the surface layers of the Arctic Ocean more salty, so that much of the icepack might not form in winter. Wouldn't that mean increased warmth, a boon to Siberians? Some Russian meteorologists questioned the scheme, defying the Communist authorities, who frowned upon any doubts cast over potential engineering triumphs. One scientist, inspecting weather records, reported that in years of less ice, there had been serious changes in precipitation around the Arctic.

Ewing and Donn's thoughts ran far beyond that. If the Arctic Ocean were free of ice, they argued, so much moisture would evaporate that heavy snows would fall all around the Arctic. The snow could build up into continental ice sheets. That was only a start. The ice buildup would withdraw water from the world's oceans, and the sea level would drop. This would block the shallow straits through which warm currents flowed into the Arctic Ocean, and it would freeze over again. Now the continental ice sheets would be deprived of evaporated moisture, and they would dwindle. The seas would rise, warm currents would spill back into the Arctic Ocean, and eventually its ice cover would melt. In a grand jumble of feedbacks, a new glacial cycle would begin.

The theory was especially interesting in view of reports that northern regions had been noticeably warming and ice was retreating. Ewing and Donn suggested that the polar ocean might become ice-free and launch us into a new ice age fairly soon—maybe even in the next few hundred years.

Scientists promptly contested what many thought was a cockamamie idea. Aside from the standard arguments that ice sheets could not grow so fast, they discovered holes in the Ewing-Donn scheme. For one, they argued, the Arctic straits were sufficiently deep so that lowering the sea level would not cut off warm currents. The pair worked to patch up the holes, and for a while many scientists found their arguments at least intriguing. In 1956 a respected U.S. Weather Bureau leader warned that "the human race is poised precariously on a thin climatic knife-edge." If the global warming trend that seemed to be under way continued, it might trigger changes with "a crucial influence on the future of the human race."[16] But ultimately the Ewing-Donn scenario won no wider acceptance than most other theories of the ice ages.

Like mistaken data, a mistaken idea can have valuable consequences. Ewing and Donn's model of the ice ages gave the public for the first time a respectable scientific backing for images of
swift, disastrous climate change. In connection with the reports on
how northern regions were growing warmer and ice was retreating, journalists had a fine time exclaiming that the climate might be
unstable. For scientists, the daring ideas provoked broad thinking
about possible mechanisms for rapid climate change in general. As
Broecker later recalled, Donn would "go around and give lectures
that made everybody mad. But in making them angry, they really
started getting into it."[17]

The new openness to thinking about rapid climate change could
get nowhere without better ways of understanding the physics of
climate. The Ewing-Donn scheme was the last influential climate
model in the grand nineteenth-century style, based on what some
called "plausibility arguments" and others dismissed as "hand-waving." The only alternative had been mathematical models—trying
to catch the pattern of global winds on a page of equations, as a
physicist might represent the orbits of planets. But in a century of
effort nobody had managed to derive a set of mathematical func-
tions whose behavior approximated that of the real atmosphere.

Starting in the late 1940s, a new way to get a handle on the problem was tried out at the University of Chicago, where Carl-Gustav
Rossby was encouraging young meteorologists to think like physicists. Dave Fultz built an actual, physical model to test the behavior
of the atmosphere. As a simulacrum of the rotating planet, his
group put a simple aluminum dishpan on a turntable, using water
to imitate the air. They represented sunlight warming the tropics
by heating the dishpan at the outer rim, and injected dye to reveal flow patterns. The results were exciting, if often mystifying.
The crude model showed something rather like the wavering polar
fronts that dominate much of the world's weather. Further experi-

ments in Chicago and in Cambridge, England, produced turntable models with something like a miniature jet stream and tiny swirling storms.

Most intriguing of all were photographs that Fultz published in 1959. His rotating dishpan showed a regular circulation pattern that resembled the real world's mid-latitude westerly winds. Rossby had explained this wind pattern as akin to the standing waves that form in swift water downstream from a rock. Stir the water in the dishpan with a pencil, and when it settled down the pattern might have flipped from a Rossby system with four standing waves to one with five—or even to an altogether different circulation pattern. This was realistic, for the circulation of the actual atmosphere does shift among quite different states; trade winds for example come and go with the seasons. Might larger shifts in the circulation pattern cause long-term climate changes?

The dishpan model was only a crude cartoon of the atmosphere. However fascinating it might be, it could not to lead to any definite conclusion about our actual planet. The real contribution of this physical model was its dramatic demonstration that certain systems might be subject to tricky instabilities. Could the real climate shift as abruptly, arbitrarily, and totally as the waves in the dishpan?

The answer would come from an altogether different way of modeling the world, one that had been tried a generation earlier and abandoned as hopeless. In 1922 Lewis Fry Richardson had proposed a complete numerical system for weather prediction. His idea was to divide up a territory into a grid of cells, each with a set of numbers for air pressure, temperature, and so forth, as measured at a given hour. He would then apply the basic physics equations that told how air would respond. He could calculate, for example, a wind speed and direction according to the difference in pressure between two adjacent cells.

The number of computations was so great that Richardson scarcely hoped his idea could lead to practical weather forecasting.

Even if someone assembled a "forecast-factory" employing tens of thousands of clerks with mechanical calculators, he doubted they would be able to compute weather faster than it actually happens. "Perhaps some day in the dim future it will be possible to advance the calculations faster than the weather advances," he wrote wistfully. "But that is a dream."[18] Still, if he could make a numerical model of a typical weather pattern, it would help show how the weather worked.

So Richardson attempted to compute how the weather over Western Europe had developed during a single eight-hour period, starting with the data for a day when there had been coordinated balloon-launchings that measured the atmosphere at various levels. The effort cost him six weeks of pencil work and ended in complete failure. At the center of Richardson's pseudo-Europe, the computed barometric pressure climbed far above anything ever observed in the real world. Taking the warning to heart, for the next quarter-century meteorologists gave up any hope of numerical modeling.

What had been hopeless with pencil and paper might possibly be made to work with digital computers. These extraordinary machines, feverishly developed during the Second World War to break enemy codes and to calculate how to explode an atomic bomb, were leaping ahead in power as the Cold War demanded more and more calculations. In the lead, energetically devising ways to simulate nuclear weapons explosions, was a brilliant and ambitious Princeton mathematician, John von Neumann. He saw parallels between weather prediction and his nuclear explosion simulations (both involved rapidly changing fluids). In 1946, von Neumann began to advocate using computers for numerical weather prediction.

The idea drew support from the U.S. Weather Bureau, the Army, and the Air Force, as well as the ubiquitous Office of Naval Research. Von Neumann told the Navy that his efforts had a dual goal: not only to predict daily weather changes, but to calculate the general circulation of the entire atmosphere, trade winds and all. That was not because he was interested in global climate change. He and his military sponsors knew they had to understand the general circulation if they were to hope to twist the climate in a given region, to their benefit or an enemy's harm.

Von Neumann asked Jule Charney, an energetic and visionary meteorologist from Rossby's group in Chicago, to spearhead the project. By 1949 Charney's team had computed air flow along a band of latitude with results that looked fairly realistic—sets of numbers that could almost be mistaken for real weather diagrams, if you didn't look too closely. For example, they could model the effects of a large mountain range on the air flow across a continent. Modeling was taking the first steps toward the computer games that would come a generation later, in which the player acts as a god: raise up a mountain range and see what happens!

The challenge was to find equations that gave plausible results without too many hours of computation. The most famous computers of the 1940s and 1950s were dead slow by comparison with a common laptop computer of later years. With thousands of glowing vacuum tubes connected by tangles of wiring, they ate up a good part of the workers' time just fixing the frequent breakdowns. Much of the remaining time was spent devising efficient and realistic mathematical approximations. These required countless hours of work, and a rare combination of mathematical ingenuity and physical insight. And that was only the beginning.

The pages of numbers produced by a computer run had to be checked against the characteristics of the actual atmosphere. This required an unprecedented number of measurements of temperature, moisture, wind speed, and so forth. Largely because of military needs, during the war and afterward networks had been established to send up thousands of balloons that radioed back measurements of the upper air. By the 1950s the weather over continental areas, up to the lower stratosphere, was being mapped well enough for comparison with the results from the rudimentary computer models.

In 1950, Charney's team solved Richardson's problem: for a chosen day they computed something roughly resembling what the weather had actually done. It took them so long to print and sort punched cards that "the calculation time for a 24-hour forecast was about 24 hours, that is, we were just able to keep pace with the weather."[19] By 1955 they had speeded up the process enough to begin issuing actual weather predictions, although it would be another decade before these could consistently beat the traditional map-reading forecasters.

These models were regional in scale, not global. But the methods and confidence developed for weather prediction inspired dreams of reaching for the ultimate prize: a model for the general circulation of the atmosphere as a whole. Norman Phillips took up the challenge. By mid-1955 he had devised a numerical model that showed a plausible jet stream and the evolution of a realistic-looking weather disturbance over a few weeks. This settled a long controversy over what processes built the pattern of circulation. For the first time scientists could see, for example, how giant eddies spinning through the atmosphere played a key role in moving energy and momentum from place to place. Phillips's model was quickly hailed as a "classic experiment"—the first true General Circulation Model (GCM).

But Phillips's model eventually exploded. As the calculations stepped forward through time, after twenty or so simulated days the pattern of flow began to look strange. By thirty days the numbers had veered off into conditions never seen on Earth. It looked like the sort of flaw that had destroyed Richardson's project back in 1922. Richardson had thought his calculation would have worked out if only he could have begun with more accurate wind data (the numbers he had used to start his computations represented weather conditions that, as a computer recalculation in the 1990s showed, were in fact impossible). But as Rossby pointed out in 1956, people routinely made decent 24-hour predictions by looking at weather maps drawn from very primitive data. "The reasons for the failure of Richardson's prognosis," the puzzled Rossby concluded, "must therefore be more fundamental."[20]

As digital computers proliferated and scientists tried them out on a variety of tasks, results often went oddly astray. The errors might indeed be caused by bad data, as Richardson had thought— "garbage in, garbage out," as computer experts were coming to understand. But the problems might also be introduced by the very nature of numerical computation. In real life a quantity like the temperature in a region varies smoothly and continuously, but the equations chopped it up into a grid. Moreover, computers do not have infinite precision, and an exact number like 5.3518277494 would be truncated to, say, 5.3518. When a set of calculations was fed into the next set thousands of times over, with the numbers truncated after each round, such tiny discrepancies could add up to a big difference. Eventually the solutions became unrealistic and "exploded." Modelers spent years devising ways to work around such artificial features in the equations and numbers.

Most scientists saw all this as a simple restriction imposed by the limited precision of computers and the limited accuracy of the network of weather instruments. A few others began to wonder whether the exquisite sensitivity of computer models might be saying something about the real world. Start two computations with exactly the same initial conditions and they must always come to precisely the same conclusion. But make the slightest change in the fifth decimal place of some initial number, and as the machine cycled through thousands of arithmetic operations, the difference might grow and grow, in the end giving a seriously different result. Of course it had long been understood that a pencil balanced on its point, for example, could fall left or right depending on the tiniest difference in initial conditions. Most scientists supposed that this kind of situation only arose under exceptionally simplified circumstances, far from the stable balance of huge and complex global systems like climate. Only a few, like Brooks, Ewing, and Donn, imagined that the entire climate system might be so delicately balanced that a relatively minor perturbation could trigger a big shift.

If that really could happen, it would be because the perturbation was amplified by feedbacks—a term and a concept that came into fashion in the 1950s. A new science of "cybernetics" was announced by mathematician Norbert Wiener, whose wartime work on automatic gun-pointing mechanisms had shown him how easily physical systems could wobble out of control. Wiener was at the Massachusetts Institute of Technology, where several groups were enthusiastically building numerical models of weather among many other things. Wiener advised meteorologists that their attempts were doomed to fail. Quoting the old nursery rhyme that told how a kingdom was lost "for want of a nail" (which caused the loss of a horseshoe of a horse bearing a knight going to a battle), he warned that "the self-amplification of small details" would foil any attempt to predict weather, to say nothing of climate.[21]

In 1961 an accident cast new light on the question. Luck in science comes to those in the right place and time with the right set of mind, and that was where Edward Lorenz stood. He was at the Massachusetts Institute of Technology, one of the new breed of professionals who were combining meteorology with mathematics. Lorenz had devised a simple computer model that produced impressive simulacra of weather patterns. One day he decided to repeat a computation in order to run it longer from a particular point. His computer worked things out to six decimal places, but to get a compact printout he had truncated the numbers, printing out only the first three digits. It was these digits that Lorenz entered back into his computer. After a simulated month or so the weather pattern diverged from the original result. A difference in the fourth

decimal place was amplified in the thousands of arithmetic operations, spreading through the computation to bring a totally new
outcome.

Lorenz was astonished. His system was supposed to represent real weather. The truncation errors in the fourth decimal place
were tiny compared with any of a hundred minor factors that
might nudge the temperature or wind speed from minute to minute. Lorenz had assumed that such variations could lead to only
slightly different solutions for the weather a few weeks ahead. Instead, storms appeared or disappeared from the forecast as if by
chance.

Lorenz did not shove this into the back of his mind, but
launched into a deep and original analysis. In 1963 he published an
investigation of the type of equations that might be used to predict
daily weather. "All the solutions are found to be unstable," he concluded. Therefore, "precise very-long-range forecasting would seem
to be non-existent."[22] Beyond a few days, or a few weeks at most,
minuscule differences in the initial conditions would dominate the
calculation. One calculation might produce a storm a week ahead,
and the next calculation, fair weather.

That did not necessarily apply to climate, which was an average
over many states of weather. Wouldn't the differences in one storm
or another balance out, on average, and leave a stable overall result?
Lorenz constructed a simple mathematical model of climate, and
ran it repeatedly through a computer with minor changes in the
initial conditions. The results varied wildly. He could not prove that
there existed a "climate" at all, in the traditional sense of a stable
long-term statistical average.

These ideas spread among climate scientists, especially at the
1965 Boulder conference on "Causes of Climate Change." Lorenz,
invited to give the opening address, explained that the slightest
change of initial conditions might bring at random a huge change

in the future climate. "Climate may or may not be deterministic," he concluded. "We shall probably never know for sure."[23] Meteorologists at the conference also pored over the new evidence that past ice ages might have been started or ended by the almost trivial astronomical shifts that Milankovitch had calculated for the Earth's orbit. Was the climate such a fundamentally unstable system that it could be tipped from one state to another by the slightest push?

That fed into discussions at the conference about the growing suspicion that the climate could change not only more easily but more swiftly than almost anyone had believed a few years earlier. Broecker, Ewing, and others presented a variety of old and new evidence that around 11,000 years ago a truly radical global climate shift had taken place: as much as 5 to 10°C in less than a thousand years. It reminded them of the sudden shifts in the rotating dishpan experiment. To be sure, melting the great ice sheets would necessarily take many thousands of years. But scientists now realized that while that was going on, the less massive atmosphere could fluctuate surprisingly easily. Summing up a consensus at the end of the conference, Revelle declared that minor and transitory changes in the past "may have sufficed to 'flip' the atmospheric circulation from one state to another."[24]

Might the ocean circulation too flip between different modes? Some were willing to question the traditional picture of a torpid circulation, unvarying over many thousands of years. Broecker, in particular, had already noticed climate jumps when he was a graduate student, comparing changes in ancient lake levels with ocean data. He had seen signs of drastic changes in ocean cores within less than a thousand years. At the Boulder meeting, Peter Weyl of Oregon State University presented an especially provocative idea. He was developing a complicated theory of the ice ages, involving how changes of saltiness might affect the formation of sea ice. It would scarcely have been noticed among the many other speculative and idiosyncratic models, except for a novel insight. Weyl pointed out that if the North Atlantic around Iceland should become less salty—for example, if melting ice sheets diluted the upper ocean layer with fresh water—the surface layer would no longer be dense enough to sink. The entire circulation that drove cold water south along the bottom could lurch to a halt. Without the vast compensating drift of tropical waters northward, a new glacial period could begin. Others since Chamberlin in the nineteenth century had speculated that the ocean circulation might be stopped if global warming somehow made northern surface waters less dense. Now explicit calculations confirmed the idea, showing that the circulation (what was coming to be called the "thermohaline circulation") was indeed precariously balanced.

Orbital changes, wind patterns, melting ice sheets, ocean circulation—everything seemed to be interacting with everything else. Not only among climate researchers, but also in other fields of science and the public mind as well, during the 1960s there was a growing recognition that the planet's environment was a hugely complicated structure. Almost any feature of the air, water, soil, or biology might be sensitive to changes in any other feature. Scientists were becoming skeptical of the traditional approach in which each expert championed a favorite hypothesis about some particular cause for climate change, blaming every shift on variations in, say, dust from volcanoes or the amount of sunlight. It seemed likely that many factors contributed together. On top of the various external influences, Lorenz and others were pointing out that complex feedbacks could make the system fluctuate under its own internal dynamics. "It is now generally accepted," wrote one authority in 1969, "that most climatic changes . . . are to be attributed to a complex of causes."[25]

This change of viewpoint, toward what some called a holistic approach, pushed scientists to change the way they worked. A plausible model for climate change could not be constructed, let alone checked against data, without information about a great many different kinds of things. It was becoming painfully clear that scientists in a variety of fields needed one another. The 1965 Boulder conference was only one of many occasions where different kinds of specialists began to interact more closely, drawing on one another's findings or, equally valuable, challenging them.

CHAPTER FOUR

A VISIBLE THREAT

In the early 1960s Walter Orr Roberts, a leading astrophysicist at the University of Colorado, noticed that something was changing in the broad and sparkling skies above Boulder. One morning, as he was talking with a journalist, he pointed out the jet airplane contrails overhead. He predicted that by mid-afternoon they would spread and thin until you couldn't tell the contrails from cirrus clouds. They did, and you couldn't. Roberts suspected that the airplanes were introducing enough additional cloudiness to affect climate in heavily traveled regions. The human impact on the atmosphere was becoming visible to the discerning naked eye.

Around the same time a University of Wisconsin meteorologist, Reid Bryson, was flying across India in a jet airplane en route to a conference. He was struck by the fact that he could not see the ground, his view blocked not by clouds, but by dust. Later he saw similar hazes in Brazil and Africa. The murk was so pervasive that local meteorologists had taken it for granted and failed to study it. But Bryson realized that the haze was not some timeless natural feature of the tropics: he was seeing smoke from fields set on fire by the growing population of slash-and-burn farmers, and dust from overgrazed lands turning to desert. The effects seemed large enough to affect the climate of the entire planet. Together with pollution from industry, the smoke and dust might block enough sunlight to significantly cool the surface. Bryson wrote that he "would be pleased to be proved wrong. It is too important a problem to entrust to a half-dozen part-time investigators."[1]

In truth scarcely anyone was studying how aerosols—microscopic airborne particles—might influence global climate. Simple physics theory suggested that aerosols should scatter radiation from the Sun back into space, cooling the Earth. Meteorologists had long suspected that dust from volcanic eruptions did just that. Aerosols not only intercepted sunlight, but might also affect climate by influencing clouds. Not only jet airplane exhaust but all sorts of human emissions might raise cloudiness, adding to the interference with sunlight. Research early in the century had shown that clouds can only form where there are enough "cloud condensation nuclei," tiny particles that give a surface for the water droplets to condense around. Interest redoubled in the 1950s, when "seeding" clouds with silver iodide smoke in hope of making rain became a widespread commercial enterprise. It was controversial, for as soon as some community attempted to bring rain on itself, people downwind would hire lawyers to argue that they had been robbed of their precipitation.

Studies of cloud seeding with special chemicals in a small locality could not tell much about the global impact of the particles emitted by human industry and agriculture. But scarcely anyone attempted to attack the global problem. Measurements and theory seemed too intractably difficult to be worth the effort. The few people who specialized in aerosols were occupied with practical problems such as their impact on public and occupational health. Scientists were starting to study local air pollution, driven by rising public dismay over urban smogs. Chemists were drawn in during the 1950s to analyze the smog of Los Angeles, which turned out to be a fascinating (and sometimes lethal) mixture of chemicals as well as particles. Meanwhile other aerosol experts worked on industrial processes such as clean rooms for manufacturing electronics, and still others investigated military problems such as the way particles scattered laser light. These researchers had only occasional contact with colleagues who studied different aerosol problems, still less with anyone in other fields of science that might relate to climate.

One respected climate student who did attack the question was J. Murray Mitchell, Jr., inspired by a new kind of aerosol. Studies of fallout from nuclear bomb tests showed that fine dust injected into the stratosphere would linger there for a few years, but would not cross from one hemisphere to the other. With that in mind, Mitchell pored over global temperature statistics and put them alongside the record of notable volcanic eruptions. In 1961 he announced that large eruptions caused a significant part of the irregular annual variations in average temperature in a given hemisphere. But he could not find a connection between volcanic eruptions and longer-term trends. In particular, the warming of the globe in the first half of the century did not match any slackening of eruptions.

In January 1961, on a snowy and unusually cold day in New York City, Mitchell told a meeting of meteorologists that while global temperatures had indeed risen until about 1940, more recently temperatures had been falling. There was so much random variation from place to place and from year to year that it was not until the late 1950s that the reversal could emerge clearly from the measurements. In any case, nobody had made a really convincing calculation of average temperatures covering most of the globe. But now Mitchell had made a plausible attempt at the exacting calculation, and he reported a clear downturn. None of the available theories of climate change seemed able to explain the recent cooling, and Mitchell could only conclude that it was "a curious enigma."[2] (See Figure 2, p. 122.)

Through the 1960s, meteorologists continued to report relatively cool average global temperatures. People will always give special attention to the weather that they see when they walk out their doors, and what they saw made them doubt that global warming was at hand. Experts who had suspected that greenhouse effect warming was on the way began to have doubts. Yet few reverted to the traditional comfortable view that climate was regulated in a stable natural balance, immune to human intervention.

That sort of view was dwindling almost everywhere. The first Earth Day, held in 1970, marked the emergence of environmentalism into full public awareness and direct political action. To many environmentalists, almost any new technology looked dangerous. Where once people had held utopian hopes for the ways humanity would modify the environment, now such "interference" began to seem ignorant, reckless, and perhaps wicked. In every democratic industrial nation, citizens pressed their government to enact environmental protection laws. Governments gave way, taking steps to reduce smog, clean up water supplies, and the like.

Like everyone else, scientists were affected by the new attitudes. They were increasingly curious to know how human activities could be warping entire geophysical systems. Reciprocally, the scientists' concepts and findings were essential in persuading the public that serious environmental harm was at hand. Contemplating the relationship between science and society, some people would say that public opinion responded intelligently to new scientific facts. Others would say that the judgment of scientists bent under the pressure of the mass prejudices of the day. Both of these views go too far in separating scientific and popular thought. In regions like North America and Europe, where the public was better informed and educated, the views of scientists and public evolved together.

A good example of the coevolution of scientific and public views began with Roberts's tentative observation, well in advance of solid scientific inquiry, that jet aircraft might increase cloudiness. The *New York Times* made it a front-page story (Sept. 23, 1963), telling its readers that jets "might be altering the climate subtly along major air routes." Such minor speculations became more serious in the mid-1960s, when the U.S. government announced plans to build a fleet of supersonic transport airplanes. Hundreds of flights a year would inject water vapor and other exhaust into the high, thin stratosphere, where natural aerosols were rare and any new chemical might linger for years. Public opposition grew swiftly. The chief objections (which drove Congress to cancel the project in 1971) were complaints that the fleet would be intolerably noisy and waste taxpayers' money. But the opposition was strengthened by scientists' warnings that emissions from a supersonic fleet could damage the atmosphere.

In 1970 the worries about supersonic transports and a number of other environmental issues inspired some policy entrepreneurs to organize a ground-breaking "Study of Critical Environmental Problems." Meeting at the Massachusetts Institute of Technology, some 40 experts in a variety of fields deliberated for a month over pollution of the air and oceans, the advance of deserts, and diverse other harms caused by humans. Among other things, they reported that, just as Roberts had suspected, cirrus clouds had increased over the United States in the past few decades. Adding a fleet of supersonic transports might alter the stratosphere as severely as a volcanic eruption. Or maybe not: the experts admitted that a calculation of actual effects was far beyond their reach. In their concluding conference report, as the first item in a list of potential problems, the scientists pointed to the global rise of CO_2. Here too effects were beyond their power to calculate. So the study could only conclude that the risk of global warming was "so serious that much more must be learned about future trends of climate change."[3] It was the old heartfelt maxim, More Money Should Be Spent on Research. The possibility of warming was so uncertain that nobody thought of recommending actual action to restrict emissions of the gas.

All but one of the participants in the study were residents of the United States, and some felt that the matter demanded a more multinational approach. This led to a comprehensive gathering of experts from 14 nations, held in Stockholm in 1971, funded by an assortment of private and government sources. It was the first major conference to focus entirely on a "Study of Man's Impact on Climate." Exhaustive discussions brought no consensus on what was likely to happen, but all agreed that serious changes were possible. The widely read report concluded with a ringing call for attention to the dangers of humanity's emissions of particle pollutants and greenhouse gases. The climate could shift dangerously "in the next hundred years," the scientists declared, "as a result of man's activities." These sober experts were adopting and endorsing the new attitudes that drove the environmentalist movement. The report's epigraph was a Sanskrit prayer: "O Mother Earth . . . pardon me for trampling on you."[4]

Weather disasters in the news reinforced the somber new mood. In 1972 a drought ravaged crops in the Soviet Union, disrupting world grain markets, and the Indian monsoon failed. In the United States the Midwest was struck by droughts severe enough to show up repeatedly on the front pages of newspapers and on television news programs. Most dramatic of all, years of drought in the African Sahel reached an appalling peak, starving millions, killing hundreds of thousands, and bringing on mass migrations. Television and magazine pictures of sun-blasted fields and emaciated refugees brought home just what climate change could signify for all of us.

It was not only the public mood that prepared climate scientists to consider the possibility of rapid and destructive climate change. They were also impressed by new data about past climates. Some of the strongest came from the University of Wisconsin, where Bryson had gathered a group to take a new, interdisciplinary look at climate. The group included an anthropologist who studied the native American cultures of the Midwest. From bones and pollen the scientists deduced that a disastrous drought had struck the region in the 1200s—the very period when the civilization of the Mound Builders had collapsed. It was already known that around that time a great drought had ravaged the Anasazi culture in the Southwest (the evidence was constricted tree rings in ancient logs from their dwellings). Compared to the drought of the 1200s, the ruinous Dust Bowl of the 1930s had been mild and temporary. A variety of historical evidence hinted that the climate shift had been world-wide. And there seemed to have been distinct starting and ending points. By the mid 1960s, Bryson concluded that "climatic changes do not come about by slow, gradual change, but rather by apparently discrete 'jumps' from one [atmospheric] circulation regime to another."[5]

Next, Bryson's group reviewed radiocarbon dates of pollen from around the end of the last ice age. In 1968 they reported a rapid shift in the mix of tree species that occurred around 10,500 years ago. Up to this point, when climate scientists spoke of "rapid" change they had meant something happening in as little as a thousand years. Bryson and his collaborators saw changes within a century. Looking at hundreds of radiocarbon dates spanning the past dozen millennia, the group arrived at a disturbing general conclusion. Periods of "quasi-stable" climate ended in catastrophic "discontinuities" when "dramatic climate change occurred in a century or two at most."[6]

To be sure, it did not require global climate change to transform some particular forest—strictly local events could do that. Many experts continued to feel it was pure speculation to imagine that the entire world's climate could change in less than a thousand years or so. If Bryson and some other climate scientists were now willing to announce that their data showed catastrophes, that was partly because the general mood of the times allowed it. At the same time, such announcements, picked up and disseminated by science journalists, were helping to shift the general mood. Scarcely any popular article on climate in the 1970s lacked a Bryson quote or at least a mention of his ideas.

Bryson's ideas could not be brushed aside, for corroborating evidence was showing up from the ends of the Earth. The first reports came from Greenland, whose frozen wastes had already played an important role in the nineteenth-century controversy over the ice ages. After a few geologists had dared to postulate the existence, in the distant past, of kilometer-thick masses of ice, astonished explorers of Greenland had found such a thing beneath their skis. Built up layer by layer over many centuries, the ice sheet carried frozen within it a record of the past.

The first attempts to exploit the ice record had begun during the 1957–58 International Geophysical Year, when a few scientists had traveled to Camp Century, a military installation high on the Greenland icecap. The daunting logistics were handled by the United States government, eager to master the Arctic regions that lay on the shortest air routes to the Soviet Union. In 1961 a specially adapted drill brought up cores of ice 5 inches in diameter in segments several feet long. This was no small feat in a land where removing your gloves for a few minutes to adjust something might cost you the skin on your fingertips, if not entire fingers.

Ice drillers began to form their own little international community. The divergent interests of people from different nations made for long and occasionally painful negotiations. But the trouble of cooperating was worth it for bringing in a variety of expertise— and a variety of agencies that might grant funds. Devising ingenious equipment, scientists managed to transport it by the ton to the ends of the earth. Costly drill heads might get irretrievably stuck a mile down, but engineers went back to their drawing-boards, team leaders contrived to get more funds, and the work slowly pushed on.[7] After five years of effort, the Camp Century drill reached bedrock at a depth of some 1.4 kilometers ($\frac{7}{8}$ of a mile), bringing up ice as much as 100,000 years old. Two years later, in 1968, another long core of ancient ice was retrieved from a site even colder and more remote, Antarctica.

Much could be read from the cores. For example, individual layers with a lot of acidic dust pointed to past volcanic eruptions. Larger amounts of mineral dust turned up deep in the ice, evidence that during the last ice age the world climate had been windier, with storms carrying dust clear from China. But the greatest hopes centered on bubbles in the ice. By good fortune there was this one thing on the planet that preserved ancient air intact, a million tiny time capsules. However, for a long time nobody could figure out how to extract and measure the air reliably.

In the early years the most useful work was done from the ice itself, using a method conceived back in 1954 by an ingenious Danish glaciologist, Willi Dansgaard. He showed that the ratio of oxygen isotopes (O^{18}/O^{16}) in the ice measured the temperature of the clouds at the time the snow had fallen—the warmer the air, the more of the heavy isotope got into the ice crystals. It was an exhilarating day for the researchers at Camp Century, who had been making measurements down each cylinder of ice as it was pulled up from the borehole, when they saw the isotope ratios change and realized they had reached the last ice age. The preliminary study of the ice cores, published in 1969, showed variations that indicated changes of perhaps 10°C. Comparison of Greenland and Antarctic cores showed the climate changes were truly global, coming at essentially the same time in both hemispheres. That immediately tossed into the wastebasket some old theories of the ice ages that relied on merely regional circumstances.

Study of ice cores confirmed a detail that Wally Broecker had already noticed in deep-sea sediments: the glacial cycle followed a sawtooth curve. In each cycle a spurt of rapid warming was followed by a more gradual, irregular descent back into the cold over many tens of thousands of years. Warm intervals like the past few thousand years normally did not last long. Beyond such fascinating hints, however, the Greenland ice cores could say little about long-term cycles. For the ice at great depths flowed like tar, confusing the record. Despite the arduous efforts of the ice drillers, in the 1970s the most reliable data were still coming from deep-sea cores.

That work too was strenuous and hazardous, manhandling long wet pipes on a heaving deck. Oceanographers (like ice drillers) lived close together for weeks or months at a time under Spartan conditions, far from their families. The teams might function smoothly—or not. Either way, the scientists labored long hours, for the problems were stimulating, the results could be exciting, and dedication to work seemed normal with everyone around them doing the same.

To make it all worthwhile, the ocean-bed drillers had to draw on all their knowledge and luck to find the right places to sample. In these few places layers of silt had built up on the ocean bed unusually swiftly and steadily, and without disturbance. During the 1960s engineers worked out drilling techniques to extract the continuous hundred-meter cores of clay that Cesare Emiliani had been asking for since the 1950s. Meanwhile scientists devised clever methods for extracting from the layers data that could shed light on past climates.

The most prominent feature was a 100,000-year cycle, so strong that it had to be a key to the entire climate puzzle. This long-term cycle had been tentatively identified in several earlier studies of seabed cores. Corroboration came from a wholly different type of record. In a brick-clay quarry in Czechoslovakia, George Kukla noticed how wind-blown dust had built up deep layers of soil (what geologists call "loess"). The multiple advances and retreats of ice sheets were visible to the naked eye in the colored bands of different types of loess. It was one of the few cases in this story where traditional field geology, tramping around with your eyes open, paid a big dividend. Analysis showed that here too, at the opposite end of the world from some of the deep-sea drilling sites, the 100,000-year cycle stood out.

But nobody could be entirely sure; radiocarbon decayed too rapidly to give dates going back more than a few tens of thousands of years. A deeper timescale could only be estimated by measuring lengths down a core, and it was uncertain whether the sediments were laid down at a uniform rate. In 1973 Nicholas (Nick) Shackleton nailed down the dates with a new technique that used radioactive potassium, which decays much more slowly than radiocarbon. He applied the timescale to a splendid deep-sea core pulled from the Indian Ocean, the famous core *Vema* 28–238 (named after Lamont's oceanographic research vessel). This core reached back over a million years, carrying a huge stretch of precise data, and Shackleton analyzed it ingeniously. When he showed his graph to a roomful of climate scientists, a spontaneous cheer went up.

Shackleton's work, confirmed by other long cores, at last proved beyond doubt what Emiliani had stoutly maintained: there had been not four major ice ages, but dozens. A sophisticated numerical analysis of measurements of the cores found an entire set of favored frequencies. The ice sheets had waxed and waned in a complex rhythm set by cycles with lengths roughly 20,000 and 40,000 years long, as well as the dominant 100,000 years. These numbers were in approximate agreement with types of orbital variations in Milankovitch's original calculations. Particularly impressive was a high-precision 1976 study of Indian Ocean cores that split the 20,000-year cycle into a close pair of cycles with lengths of 19,000 and 23,000 years. That was exactly what the best new astronomical calculations predicted for the "precession of the equinoxes," a wobbling of the Earth's axis. These studies left most scientists convinced that orbital variations were central to long-term climate change.

There remained the old objection that the subtle changes in the amount of sunlight reaching the Earth were far too small to affect climate. Worse, the record was dominated by the 100,000-year cycle, which brought particularly tiny variations of sunlight (caused by a minor change in "eccentricity," a slight stretching of the Earth's path around the sun away from a perfect circle). The only reasonable explanation, as Shackleton and others immediately understood, must be that feedbacks were amplifying the changes. Presumably the feedbacks involved other natural cycles that resonated at roughly the same time-scale. The orbital variations only served as a "pacemaker" that pinned down the exact timing of internally driven feedback cycles. In Milankovitch's day, most climate scientists had thought such a thing unlikely. By the mid-1970s scientists were seeing feedback cycles everywhere, poised to react with hair-trigger sensitivity to external influences.

What were the natural cycles that fell into step with the shifts of sunlight? The most obvious suspect was the continental ice sheets. It took many thousands of years for snowfall to build up until the ice began to flow outward. A related suspect was the solid crust of the Earth. On a geological scale it was not truly solid, but sluggishly sagged where the great masses of ice weighed it down, and sluggishly rebounded when the ice melted. (Scandinavia, relieved of its icy burden some twenty thousand years ago, is still rising a few millimeters a year.) Since the 1950s scientists had speculated that the timing of glacial periods might be set by these slow plastic flows, the spreading of ice and the warping of crustal rock. During the 1970s a number of scientists invented elaborate numerical models that suggested how 100,000-year cycles might be driven by feedbacks among ice buildup and flow, along with the associated movements of the Earth's crust, changes in reflection of sunlight, and the rise and fall of sea level. The models were obviously speculative, for nobody had reliable equations or data for such processes. But it did seem possible that some kind of natural feedback systems could amplify the weak Milankovitch sunlight changes (and perhaps other variations too?) into full-blown ice ages.

Whatever caused it, the curve of ancient temperatures followed the complex pattern of astronomical cycles with uncanny accuracy. It was natural to continue the calculations and extrapolate the curve forward in time. The predicted curve headed downward for the next 20,000 years or so. Emiliani, Kukla, Shackleton, and other specialists concluded that the Earth was gradually heading into a new ice age.

Or perhaps not so gradually? In 1972 Murray Mitchell, the respected climate expert from the U.S. Weather Bureau, said that in place of the old view of "a grand, rhythmic cycle," new evidence was revealing a "much more rapid and irregular succession" in which the Earth "can swing between glacial and interglacial conditions in a surprisingly short span of millennia (some would say centuries)."[8] Evidence for these swings, in particular the Younger Dryas oscillation around 12,000 years ago, had turned up in radiocarbon studies of ancient glacier moraines, fossil shorelines, lake levels, shells in the most undisturbed deep-sea cores, and more. The most convincing evidence came from the Greenland core drilled by Dansgaard's group of Danes and Americans. Mixed in with the gradual cycles were what Dansgaard called "spectacular" shifts, lasting perhaps as little as a century or two—including, once again, the Younger Dryas.[9] Perhaps that was an illusion, for the flow of ice at great depths muddled the record. Or perhaps back then the climate around the North Atlantic, if not around the entire globe, really had changed drastically.

If scientists were increasingly willing to consider that global climate might change greatly in the space of a century, it was because different pieces of the puzzle seemed to fit together better and better. Sudden lurches had persistently turned up in the rudimentary weather and climate models developed in the 1950s and 1960s, built from rotating dishpans or from sets of simple equations run through computers. Scientists could have dismissed these models as too crude to say anything reliable—but the historical data showed that the notion of radical climate instability was not absurd after all. And the researchers could have dismissed the jumps in the data as artifacts due to merely regional changes or simple errors—but the models showed that such jumps were physically plausible.

Even equations that scientists were developing to model continental ice sheets offered disturbing prospects of abrupt changes. In 1962 John Hollin had argued that the great volumes of ice in Antarctica, piled up kilometers high and pushing slowly toward the ocean, were held in place by their fringes. These edge sheets were pinned in turn at the marginal "grounding line" where they rested on the ocean floor. A rise in sea level could float an ice sheet up off the floor, releasing the entire stupendous mass behind it to flow more rapidly into the sea. The idea was picked up by Alex Wilson, who pointed to the spectacle of a "surge." Glaciologists had long been fascinated by the way a mountain glacier might suddenly give up its usual slow creeping, to race forward at hundreds of meters a day. They figured this happened when the pressure at the bottom melted ice so that water lubricated the flow. As the ice began to move, friction melted more water and the flow accelerated. Could the ice in Antarctica become unstable in this fashion?

If so, the consequences sketched by Wilson would be appalling. As the ice surged into the sea, the world's sea coasts would flood. But that would be almost the least of humanity's problems. Immense sheets of ice would float across the southern oceans, cooling the world by reflecting sunlight, bringing on a new ice age.

Through the 1960s few scientists gave much credence to these ideas. The ice that covered most of Antarctica, in places over four kilometers thick, seemed firmly grounded on the continent's bedrock. But around 1970, a glaciologist at Ohio State University, J. H. Mercer, drew attention to the West Antarctic Ice Sheet. This is a smaller (but still enormous) mass of ice, separated by a mountain range from the bulk of the continent. Mercer argued that this mass was held back in an especially delicate balance by the ice shelves floating at its rim. The shelves might disintegrate under a slight warming. If the West Antarctic Ice Sheet was released and slid into the oceans, the sea level would rise as much as 5 meters (16 feet), forcing the abandonment of many great cities. This could happen rapidly, Mercer thought, perhaps within the next 40 years.

Other glaciologists looked into the idea, building models of ice movements using data from adventurous survey expeditions that had traversed parts of Antarctica during the International Geophysical Year and on later occasions. They found that the West Antarctic Ice Sheet might indeed be unstable. It was "entirely possible," one author concluded in 1974, that the sheet was already now starting to surge forward.[10] Most climate specialists and geologists felt that these models were highly speculative. It seemed scarcely possible that the West Antarctic sheet could disintegrate in less than a few centuries. But a surge that dumped a fifth of a continent of ice into the oceans over the next few centuries would be no small thing, and the models were too inadequate to rule out an even faster collapse.

If the climate experts of the 1970s seem to have been a bit preoccupied with ice, that fitted their training and interests. For a century their field had concerned itself above all with the ice ages. Their techniques, from pollen studies to ice core drilling, were devoted to measuring the swings between warm and glacial epochs. Home at their desks, they occupied themselves with figuring how glacial climates had differed from the present, and attacked the grand challenge of explaining what might cause the swings. Now that they were beginning to turn their attention from the past to the future, the most natural meaning to attach to "climate change" was the next swing into cold.

In 1972 a group of leading glacial-epoch experts met at Brown University to discuss how and when the present warm interglacial period might end. Reviewing the Greenland ice cores, Emiliani's foraminifera, and other field evidence, they agreed that interglacial periods tended to be short and to end abruptly. A large majority further agreed that extrapolating the Milankovitch curves into the future showed that "the natural end of our warm epoch is undoubtedly near." They noted that weather records showed the world had not been warming since 1940. Arctic regions, which seemed especially sensitive to such trends, showed signs of cooling. The scientists were far from reaching complete agreement, with some insisting that any cooling might be counteracted by greenhouse effect warming, or by other, unknown factors. But the majority concurred in a statement that serious cooling "must be expected within the next few millennia or even centuries."[11]

Several members of the study group wrote a letter to President Richard Nixon, calling on the government to support intensified studies. That was one example of a general movement during the 1970s. Scientifically trained people were making contact with policy elites to address the planet's environmental future.

The study group suspected that natural cycles were not the greatest hazard. While some worried about global warming from CO_2, others now saw a greater risk of global cooling. Would the drop in temperature that the Milankovitch schedule predicted in future millennia be accelerated by smoke and dust from human agriculture and industry? Bryson, for one, argued with increasing conviction that such a rapid cooling was only too likely.

As Bryson had noticed while flying over the tropics, entire regions could be hazed over for months at a time. Already in 1958, one expert had remarked that "there can no longer be any sharp division between polluted and unpolluted atmospheres."[12] Most meteorologists, however, were slow to notice the spread of pollution beyond cities. Scientists concerned with pollution particles mostly studied dusts that fell out of the atmosphere within a few days, not the microscopic particles that lingered much longer. Better understanding came only after people studying smog set up a network of stations that regularly monitored the atmosphere's turbidity (haziness). In 1967, two scientists at the National Center for Air Pollution Control in Cincinnati reported a gradual increase in the general turbidity over regions spanning a thousand kilometers. Further checks of the record of turbidity turned up increases even in remote areas like Hawaii and the North and South Poles. Could humanity's emissions be affecting the global climate not in some abstract future, but right now?

These studies contributed to, and at the same time responded to, the general shift of public viewpoint. The world's oceans and air could no longer be seen as a virtually infinite dumping-ground that could safely absorb any and all emissions. Concern grew following reports that the air over the North Atlantic was twice as dirty in the late 1960s as it had been in the 1910s, suggesting that the natural processes that washed aerosols out of the atmosphere could not keep up with human emissions. As an item in the back pages of the *New York Times* reported (Oct. 18, 1970), "This is disturbing news for those weather experts who fear that air pollution, if it continues unchecked, will seriously affect the climate and perhaps bring a new ice age."

But how much of the haze was really caused by humans? In 1969 Mitchell pushed ahead with his statistical studies of temperatures and volcanoes. He calculated that about two-thirds of the cooling seen in the Northern Hemisphere since 1940 was due to a few volcanic eruptions. He concluded that "man has been playing a very poor second fiddle to nature as a dust factory."[13] Other respected climatologists agreed that one could look to volcanic dust to account for a substantial part, if not all, of the temperature variations in the last century or so. They could not agree on how strong the effect of volcanic eruptions was, and how strong the effect of human pollution. They could only admit that these questions had been overlooked for too long, and deserved sustained scrutiny.

In 1971, S. Ichtiaque Rasool and Stephen Schneider entered the discussion with a pioneering numerical computation. (This was the first atmospheric science paper by Schneider, who would later become a well-known commentator on global warming.) Developing ideas suggested by Mitchell and others, Rasool and Schneider explained how the effect of aerosols could vary. Some kinds of haze might not cool the atmosphere after all, but warm it. It depended on how much the aerosols absorbed radiation coming down from the Sun, and how much they trapped heat radiation rising up from the Earth's surface. Rasool and Schneider's calculation gave cooling as the most likely result. Estimating that dust in the global atmosphere might have doubled already during the century, and might double again in the next fifty years, they figured that this might seriously cool the planet, as much as 3.5°C. Rasool and Schneider also believed the cooling would not be counteracted by the greenhouse effect, since according to their model there would be little warming even if a lot of CO_2 were added. The dip due to aerosols, they exclaimed, "could be sufficient to trigger an ice age!"[14] In fact their equations and data were rudimentary, and critics swiftly pointed out crippling flaws. But if they were wrong, what was the correct interpretation?

Beyond the direct effects of aerosols as they absorbed or scattered radiation, there remained an even greater mystery: how did particles help create particular types of clouds? And beyond that loomed an equally enigmatic problem: how did a given type of cloud reflect radiation coming from above or below, to cool the Earth or perhaps warm it? Simplified calculations turned up all sorts of subtle and complex influences. The one sure thing was that aerosols could make a difference to climate, perhaps a big difference.

Many people now wanted to know just how big a difference aerosols or CO_2 or other human products could make. To answer that grave question, scientists needed something better than crude hand-waving models. They turned to mathematical calculations. In 1963 Fritz Möller, building on the pioneering work of Gilbert Plass (Chapter 2), had produced a model for how radiation moves up and down a column of typical air (the shorthand term was a "one-dimensional global-average" model). His key assumption was that the humidity of the entire atmosphere should increase with increasing temperature. To put this into the calculations he held the relative humidity constant, which was just what Arrhenius had done in his pioneering calculations back in 1896 (Chapter 1). Möller got the same amplifying feedback that Arrhenius had found. As the temperature rose, more water vapor would remain in the air, adding its share to the greenhouse effect.

When he finished his calculation, Möller was astounded by the result. Under some reasonable assumptions, doubling CO_2 in the atmosphere could bring a temperature rise of 10°C—or perhaps even higher, for the mathematics would allow an arbitrarily high rise. More and more water would evaporate from the oceans until the atmosphere filled with steam! Möller found this result so implausible that he doubted the whole theory. Indeed, his method was later shown to be fatally flawed. But most research begins with flawed theories, which prompt people to make better ones. Some scientists found Möller's calculation fascinating. Was the mathematics trying to tell us something truly important? It was a disturbing discovery that a simple calculation (whatever problems it might have in detail) could produce a catastrophic outcome. That was one stimulus for taking up the challenging job of building full-scale computer models.

Through the 1960s, however, computer simulations of the entire general circulation of the atmosphere remained primitive. Even if the calculations ran for weeks on end, the results looked only roughly like the present climate. That could scarcely tell people how things might change under a subtle influence. So some climate scientists tried using computers in a less expensive and arduous way. They built highly simplified models that could work out rough numbers in a few minutes. While some scientists gave these simple models no credence, others felt that they were valuable "educational toys"—a helpful starting point for testing assumptions and for identifying spots where future work could be fruitful.[15]

The most important simple model of climate change was built by Mikhail Budyko in Leningrad. He was drawn to the issue by the grandiose proposals that had concerned Soviet climatologists since the 1950s. If they did not divert rivers from Siberia, how about warming the Arctic by spreading soot over snow and ice to absorb sunlight? Studying historical data on how the snow cover in a region connected with temperature, Budyko found a dramatic interdependence.

To pin down his ideas, around 1968 Budyko constructed a highly simplified set of equations. They represented the heat balance of the Earth as a whole, summing up over all latitudes the incoming and outgoing radiation: an "energy budget" model. When he plugged plausible numbers into his equations, he found that for a planet under given conditions—that is, a particular atmosphere and a particular amount of radiation from the Sun—more than one state of glaciation was possible. If the planet had arrived at the present state after cooling down from a warmer climate, the reflection of sunlight ("albedo") back into space by the dark sea and soil would be relatively low. So the planet could remain entirely free of ice. If it had come to the present by warming up from an ice age, keeping some snow and ice that reflected sunlight, it could retain its chilly ice caps.

Under present conditions, the Earth's climate looked stable in Budyko's model. But not too far above present temperatures and albedo, the equations reached a critical point. The global temperature would shoot up as the ice melted away entirely, uncovering soil and water. That would give a uniformly and enduringly warm planet with high ocean levels, as seen in the time of the dinosaurs. And if the temperature dropped not too far below present conditions, the equations hit another critical point. Here temperature could drop precipitously as more and more water froze, until the

Earth reached a stable state of total glaciation: the oceans entirely frozen over, the Earth transformed permanently into a gleaming ball of ice! Budyko thought it possible that our era was one of "coming climatic catastrophe . . . higher forms of organic life on our planet may be exterminated."[16]

Other analysts were on the same trail, independently of Budyko's work in Leningrad—communications were sporadic across the Cold War frontiers. What finally caught full attention was an energy-budget model published in 1969 by William Sellers at the University of Arizona. His model was still "relatively crude," as Sellers admitted (adding that this was unfortunately "true of all present models"), but it was straightforward and elegant. Climatologists were impressed to see that although Sellers used equations quite different from Budyko's, his model too could approximately reproduce the present climate—and it also showed a cataclysmic sensitivity to small changes. If the energy received from the Sun declined by two percent or so, whether because of solar variations or increased dust in the atmosphere, Sellers thought it might bring on another ice age. Beyond that, Budyko's nightmare of a totally ice-covered Earth seemed truly possible. At the other extreme, Sellers suggested, "man's increasing industrial activities may eventually lead to a global climate much warmer than today."[17]

Did the Budyko-Sellers catastrophes reflect real properties of the global climate system? That was a matter of brisk debate. In the early 1970s some scientists did find it plausible that feedbacks could build up a continental ice sheet more rapidly than had been supposed. The opposite extreme—a self-sustaining heating—might be even more catastrophic. For there was new evidence that another calculation from a few equations, Möller's preposterous runaway greenhouse, represented something real.

A planet is not a lump in the laboratory that scientists can subject to different pressures and radiations, comparing how it reacts to this or that. We have only one Earth, and that makes climate science difficult. To be sure, we can learn a lot by studying how past climates were different from the present one. But these are limited comparisons—different breeds of cat, but still cats. Fortunately our solar system contains wholly other species, planets with radically different atmospheres.

Radio observations of the planet Venus, published in 1958, indicated that the surface temperature was amazingly hot, around the melting point of lead. Why would a planet about the same size as the Earth, and not all that much closer to the Sun, be so drastically different? In 1960 a young doctoral student, Carl Sagan, took up the problem and got a solution that made his name known among astronomers. Using what he later recalled were embarrassingly crude methods, taking data from tables designed for steam-boiler engineering, he showed that the greenhouse effect could make Venus a furnace. He thought this was mainly due to water vapor, in a self-perpetuating process. The surface was so hot that whatever water the planet possessed went into the atmosphere as vapor, helping maintain the extreme greenhouse effect condition. It was later found that Venus's atmosphere has little water. Sagan was wrong—another of those scientific mistakes that usefully stimulated further work.

A few researchers pursued the feedback between temperature and water vapor in simple systems of equations, with strange results. In 1969 Andrew Ingersoll reported "singularities," mathematical points where the numbers went out of bounds. He pointed at CO_2 as the main culprit. On our planet most of the carbon has always been locked up in minerals and buried in sediments. The surface of Venus, by contrast, was so hot and dry that carbon-bearing compounds evaporated rather than remaining in the rocks. Thus its atmosphere was filled with a huge quantity of the greenhouse gas. Perhaps Venus had once enjoyed a climate of the sort hospitable to life, but as water had gradually evaporated into the warming atmosphere, followed by CO_2, the planet had fallen into its present hellish state. (The question of whether Venus once had a more Earthlike climate remains unresolved today.) According to one calculation, the Earth would need to be only a little warmer for enough water to evaporate to tilt the balance here as well. If our planet had been formed only 6 percent closer to the Sun, the authors announced, "it may also have become a hot and sterile planet." This was published in 1969, the same time as the work of Budyko and Sellers.[18]

And then there was Mars. In 1971 the spacecraft Mariner 9, a marvelous jewel of engineering, settled into orbit around Mars and saw . . . nothing. A great dust storm was shrouding the entire planet. Such storms are rare for Mars, and this one was no misfortune for the observers, but great good luck. They immediately saw that the dust had profoundly altered the Martian climate, absorbing sunlight and heating the planet by tens of degrees. The dust settled after a few months, but its lesson was clear. Haze could indeed warm an atmosphere. More generally, anyone studying the climate of *any* planet would have to take dust very seriously. Moreover, it seemed that the temporary warming had reinforced a pattern of winds that had kept the dust stirred up. It was a striking demonstration that feedbacks in a planet's atmospheric system could flip weather patterns into a drastically different state. That was no longer speculation but an actual event in full view of scientists—"the only global climatic change whose cause is known that man has ever scientifically observed."[19]

Before Mariner arrived at Mars, Sagan had made a bold prediction. He suggested that the Red Planet's atmosphere could settle in either of two stable climate states. Besides the current age of frigid desiccation, there was another possible state, more clement, which might even support life. The prediction was validated by crisp images of the surface that Mariner beamed home after the dust cleared. The canals some astronomers had once imagined were nowhere to be seen, but geologists did see strong signs that vast water floods had ripped the planet in the far past. Calculations by Sagan and his collaborators now suggested that the planet's climate system was balanced so that it could have been flipped from one state to the other and back by relatively minor changes.

Such theories of change now dominated thinking about climate on Earth as well. People no longer imagined the climate was permanently stable. It was not the weight of any single piece of evidence that was convincing, but the accumulation of evidence from different, independent fields. Mars and Venus, increased haze and jet contrail clouds, catastrophic droughts, fluctuating layers in ice and in sea-bed clays, computer calculations of planetary orbits and of energy budgets and of ice sheet collapse: each told a story of climate systems prone to terrible lurches. Each story, bizarre in itself, was made plausible by the others. Most experts knew only some parts of the evidence. Most members of the public knew hardly any of it. But the main idea got around.

To be sure, no tremendous climate change had come in recent memory. But people were ready to take a longer view. During the first half of the century, violent interruptions of life by wars and economic upheavals had made long-term planning seem pointless. By the 1970s, more settled times encouraged people to think farther ahead. If human emissions were liable to change the climate in the twenty-first century, that no longer seemed too far off to worry about.

Maybe somebody should do something about it?

CHAPTER FIVE

PUBLIC WARNINGS

The politics of science came up briefly at a 1972 symposium where scientists were discussing the rising level of CO_2 in the atmosphere. Should they make a statement calling for some kind of action? "I guess I am rather conservative," one expert remarked. "I really would like to see a better integration of knowledge and better data before I would personally be willing to play a role in saying something political about this." A colleague replied: "To do nothing when the situation is changing very rapidly is not a conservative thing to do."[1]

Most scientists felt they were already doing their job by pursuing research and publishing it. Anything important would presumably be noticed by science journalists and policy-makers. For really important problems the scientists could convene a study group, perhaps under the National Academy of Sciences, and issue a report. Experts like Roger Revelle and Reid Bryson were more than willing to explain their ideas when asked, and they might even make an effort to come up with quotable phrases for reporters. They were glad to give talks on the state of climate science, or write an article for a magazine like *Scientific American*. Such efforts would reach, if not exactly the public, the small segment of the public that was well educated and interested in science.

A wider public would only take notice if something special came

along, something newsworthy. In the early 1970s, the climate offered plenty of such opportunities. The dramatic droughts and
crop failures in India, Russia, and the American Midwest, and the
deadly famine in Africa sent journalists to talk with climate scientists time and again. They reported how some scientists suspected
the weather fluctuations could be the harbinger of another ice age.
After all, many scientists now accepted that the long-term trend, extrapolating the Milankovitch curves, should be a gradual cooling.
But if that would happen in the absence of human intervention,
what might human intervention do? The leader in stirring public
anxiety, quoted more than any other expert, was Bryson. He wanted
everyone to know about the increase in smoke and dust caused by
industry, deforestation, and overgrazing. Like the smoke from a
huge volcanic eruption, he said, the "human volcano" could bring a
disastrous climate shift. The effects, he declared, "are already showing up in rather drastic ways." As rising population crashed against
the increasingly erratic weather, the world faced mass starvation.[2]

Most climate experts insisted that such ideas were by no means
scientific predictions, but only possibilities. Responsible journalists
made it clear that every expert admitted ignorance. Yet the majority
of scientists did feel, as *Time* magazine put it, that "the world's prolonged streak of exceptionally good climate has probably come to
an end—meaning that mankind will find it harder to grow food."[3]
The most common scientific viewpoint was summed up by a scientist who explained that the rise in dust pollution worked in the opposite direction from the rise in CO_2, so nobody could say whether
there would be cooling or warming. But in any case, "We are entering an era when man's effects on his climate will become dominant."[4]

Whether our emissions would speed the coming of a new ice age
or bring on greenhouse warming, the moral lesson seemed the
same. "We have broken into the places where natural energy is
stored and stolen it for our own greedy desires," a journalist de-

claimed. Some people, he said, expect that "our tampering with the delicate balances of nature" would call down "the just hand of divine judgment and retribution against materialist sinners."[5]

More optimistic people suggested that we might counter any ill effects by deliberately bending the climate to our will. If an ice age approached, we might spread soot from cargo aircraft to darken the Arctic snows, or inject sunlight-absorbing smog into the stratosphere, or even shatter the Arctic ice pack with "clean" thermonuclear explosions. Most scientists dismissed such ideas, but not because they sounded like science fiction. It seemed only too plausible that humanity could alter the climate. But the bitter fighting among communities over cloud-seeding would be as nothing compared with the conflicts that attempts to engineer global climate might bring. And our knowledge was so primitive that intervention might only make things worse.

The general public was aware of all this talk, but only vaguely. Since the nineteenth century, news media had trumpeted future threats from this or that quarter, and the public saw nothing special here. Reports of scientific findings were usually relegated to a few paragraphs on the inside pages of the better newspapers or in the science-and-culture section of news magazines, reaching only the more alert citizens. It was Nigel Calder, a respected British science journalist, who first alerted a somewhat broader public— those who watched educational shows on public television—to the threat of climate catastrophe. In 1974 he produced a two-hour feature on weather, which spent a few minutes warning of a possible "snowblitz." A sudden cooling could be set off, said Calder, by an Antarctic ice surge, or global warming, or atmospheric pollution, or just pure chance. Entire countries could be obliterated under layers of snow; billions would starve. The new ice age "could in principle start next summer, or at any rate during the next hundred years."[6]

Most experts despised such talk. They felt that the public was being led astray by a few scientists and sensation-seeking journalists. The official message from the Director-General of the United Kingdom Meteorological Office was: "no need for panic induced by the prophets of doom." Like many meteorologists, he held that "the climatic system is so robust . . . that man has still a long way to go before his influence becomes great enough to cause serious disruption."[7] The traditional belief in a benign Balance of Nature was still widely held among scientists as well as the public. Most scientists preferred just to get on with their research; after a few more decades they might know enough to deal with the question responsibly.

A few scientists took the prospects of climate catastrophe so seriously that they felt they should make a personal effort to address the public directly. Bryson wrote a book titled *Climates of Hunger*, published in 1977. He described how some primitive societies had been destroyed by sudden droughts, far worse than anything known in recent centuries, and warned that such a disaster could hit our own civilization.[8] Another climatologist who worked hard to make his ideas heard was Stephen Schneider. He and his journalist wife wrote a popularizing book, *The Genesis Strategy: Climate and Global Survival*. Insisting that climate could change more quickly and drastically than most people imagined, they advised the world to devise policies to cushion the shocks, by building a more robust agricultural system, for example. As Joseph had advised Pharaoh in the Book of Genesis, we should prepare for lean years to follow fat ones.[9]

Some scientists criticized Bryson, Schneider, and others for speaking directly to the public. The time spent writing a book and going about the country delivering public lectures was time away from doing "real" science. Besides, most scientists felt that any definite statement about climate change was premature. The whole subject was so riddled with uncertainties that it seemed unfit for presentation, in a few simplified sound bites, to the scientifically naive public. But like it or not, the issue was becoming political, at least in the narrow sense that policies were at stake. At professional meteorological conferences, debates over technical questions such as the rate of CO_2 buildup became entangled with debates over how governments should respond. Scientists began to struggle with questions far beyond their professional expertise. Should reliance on fossil fuels be reduced? How much money should be spent on averting climate change, amid the struggle to feed the world's poor? Was it proper for a scientist to speak, as a scientist, on social and economic questions?

Unable to agree even whether the world was likely to get warmer or colder, the scientists did unanimously agree that the first step must be to redouble the effort to understand how the climate system worked. Calls for research always came naturally to researchers, but from the early 1970s onward climate scientists issued such calls with increased frequency and passion. They had never seen such strong reasons to insist on their traditional principle, More Money Should Be Spent on Research.

The science of long-term climate change nevertheless remained a minor topic. A rapid rise in publications had begun around 1950, but that did not mean much, for the starting level had been negligibly small. Around 1970 the advance had stalled, and the rate of publication was now rising only sluggishly. Into the mid-1970s, well under 100 scientific papers per year were published worldwide on any aspect of the subject.[10] During those economically stagnant times, the funding for climate science in every country was generally static. One reason was that the funds were dispersed among a variety of private organizations and relatively small and weak government agencies.

One chance for improvement did come in 1970. Funding of the ocean sciences was particularly scattered, and a group had been promoting a centralized agency—a "wet NASA." With a boost from the rise of environmentalism, scientists prodded the U.S. government not only to consolidate the nation's maritime programs, but to merge them with atmospheric research. The result was a new organization, the National Oceanic and Atmospheric Administration, which was to address the planet's entire fluid envelope. From the outset, NOAA was one of the world's chief sources of funding for basic climate studies. However, the agency was created by rearranging programs without adding new money. And if NOAA had a central focus, it was in developing economically important marine resources such as fisheries. The atmospheric sciences were left mired in ambiguity. Climate scientists continued to look for an organization and dedicated funding all their own.

When a group of citizens (in this case scientists) decides that their government should do more to address some particular concern, they face a hard task. The citizens have only a limited amount of effort to spare, and officials are set in their bureaucratic ways. To accomplish anything—to bring about a new government program, for example—people must mount a concerted push. For a few years concerned citizens must hammer at the issue, informing the public and forging alliances with like-minded officials. These inside allies must form committees, draft reports, and shepherd legislation through the administration and legislature. Roadblocks will be put up by special interests that feel threatened by change, and the whole process is liable to fail from exhaustion. Typically such an effort succeeds only when it can seize a special opportunity, usually news events that distress the public and therefore catch the eye of politicians.

In the early 1970s, a few climate scientists sought such an opportunity to mount such a concerted push. They were spurred by the new calculations and data which convinced them that climate might change far sooner and more drastically than had seemed possible only a decade earlier. The media furor over droughts and other environmental problems that erupted in the early 1970s gave climate scientists a chance to take action.

Most took a traditional route: they convened study groups and prepared reports for policy-makers. For example, the U.S. National Academy of Sciences established a Committee on Climatic Variation. In 1974 the group issued a report that warned of the risk of climate shifts and recommended a national climate research plan. The administration began to draft legislation to improve the organization and funding of the field. In 1976, with the recent droughts much in mind, a Congressional committee began hearings. These were the first ever to address climate change as their main subject, starting what would become a long procession of scientists testifying that the rise of CO_2 could bring calamity. Meanwhile agency officials wrote and rewrote plans, negotiating tenaciously over who should get control of what research budget. Scientists kept up the pressure with a 1977 Academy report on "Energy and Climate," warning again that climate shocks might be in store—truly, More Money Should Be Spent on Research.

The Academy's experts were by no means prepared to go farther and recommend actual changes in the nation's energy policies. They knew climate predictions were too unreliable to support such a move. If they avoided concrete advice, they did drive home a general truth: the threat of climate change was intimately connected with energy production. As a page-one headline in the *New York Times* (July 15, 1977) summed it up, "Scientists Fear Heavy Use of Coal May Bring Adverse Shift in Climate." Officials were starting to grasp the fact that CO_2 emissions had economic implications—and therefore political ones. The oil, coal, and electrical power industries began to pay attention.

Fossil-fuel policies were already under intense scrutiny. In the 1973 "energy crisis," inconvenience and anxiety beset millions of people as Persian Gulf states withheld their oil. When President Jimmy Carter's administration proposed to shift the United States from oil to coal, politics began to overlap scientific studies of climate change. The energy crisis gave a boost to advocates of renewable energy sources, ranging from Federal solar-energy bureaucrats to anti-government environmentalists. They found the greenhouse effect useful in arguing for their cause: more power generated by windmills would mean less CO_2 emissions. In the energy debates, however, climate change was only one more weight thrown into the balance, and compared with the many economic, political, and international issues, it was far from the heaviest weight in people's minds.

Nobody of consequence proposed to regulate CO_2 emissions or make any other significant policy changes to deal directly with greenhouse gases. Academy reports and other scientific pronouncements advised that any such action would be premature, given the lack of scientific consensus. The goals of the meteorological community and its friends in the bureaucracy remained the same: more money, and better organization, for research.

The spearhead of the effort was a Climate Research Board set up by the National Academy. The Board's full-time chair, Robert M. White, was a widely admired scientist-administrator who had already served as head of the Weather Bureau and then of NOAA, as official representative to the World Meteorological Organization as well as to various international meetings—one on whaling, for example, another on desertification—and in countless other capacities. Bob White deserves notice as the most outstanding example of many people whose names are not mentioned here, but whose contributions in administration and organization were indispensable.

At last in 1978 Congress passed a National Climate Act, establishing a National Climate Program Office within NOAA. It was a step forward, but the new office had only a feeble mandate and a budget of only a few million dollars. Scientists did not get as well-coordinated a research program as they had called for. Without the backing of some unified community or organization, their movement had been impeded by the very fragmentation it sought to remedy. Any reorganization that might usurp the authority of existing research bureaucracies stood little chance. Moreover, lawmakers cared far more about the few years until the next election than about the following century.

The one solid success was at the new cabinet-level Department of Energy, established in response to the energy crisis. Hard-driving officials there seized a share of CO_2 work, and won large budget increases. Yet, in a pattern that would often be followed when the government wanted to boast of its support for environmental causes, some of the expansion in the formal budget was not new money, but only a transfer of funds that had already been available through other programs.

With the passage of the National Climate Act, the minor flurry of legislative attention ended. The program to study climate change was underfunded from the start, and the large increases won by the Department of Energy in the 1970s came to a dead halt in 1980 as Congress tried to balance the budget. Such new money as was available seemed to go more into paperwork and meetings than into actual research.

Climate scientists in every nation had found it difficult to gain access to their respective policy-makers. If they convinced their contacts among lower-level officials that there was a problem, these officials themselves had scant influence in higher reaches of government. The scientists discovered better opportunities when they turned to the international science community. Efforts by a group of nations—not just their combined money but the consensus of their prestigious scientists—might help convince a given nation's politicians to act. Besides, internationalization might offer some of the organization that was needed. Scientists and science officials concerned about climate change found their best opportunities in working with foreign colleagues. After all, for something as global as weather nobody could get far without exchanging information and ideas across national borders.

The World Meteorological Organization (WMO) and the 1957–58 International Geophysical Year (IGY) had made a good start, but they fell far short of gathering the kind of global data needed to understand the atmosphere. For example, even at the peak of the IGY there had been only one station reporting upper-level winds for a swath of the South Pacific Ocean that spanned one-seventh of the Earth's circumference. The lack of data posed insuperable problems for atmospheric scientists. Officials and scientists managed to bring the problem to the attention of U.S. President John F. Kennedy, who saw an opportunity for enhancing his administration's prestige with a bold initiative. Addressing the United Nations General Assembly in 1963, Kennedy called for a cooperative international effort aimed at better weather prediction and eventually weather control. The proposal was eagerly taken up by the WMO. In 1963, the organization launched a World Weather Watch, coordinating thousands of professionals reading weather gauges, launching balloons, analyzing satellite pictures, and so forth. The Watch has continued down to the present day as the core WMO activity. It has served weather forecasters everywhere, unimpeded by the Cold War and other international conflicts.

The WMO was a federation of government weather agencies, officials who were only loosely connected with academic scientists. The scientists had long since organized themselves in specialized societies such as the International Union of Geodesy and Geophysics, which loosely cooperated under the umbrella of an International Council of Scientific Unions (ICSU). Determined not to be left out in organizing weather research, ICSU negotiated with the WMO. In 1967 the two organizations jointly formed a Global Atmospheric Research Program (GARP). The chief aim was to improve short-term weather prediction, but climate research was included. Once the top experts on a GARP scientific committee had forged plans for a cooperative program, it was hard for the various national budget agencies to deny their scientists the funds needed to join in.

The chair of GARP's organizing committee during its crucial

formative years was a Swedish climatologist, Bert Bolin. A savvy expert on subjects ranging from weather computation to the global carbon cycle, but still more admired for his skills as a team leader and diplomat, Bolin would be a mainstay of international climate organizing efforts for the next three decades.

Climate scientists met one another in an increasing number of international scientific meetings, from cozy workshops to swarming conferences. The 1971 Stockholm "Study of Man's Impact on Climate" broke new ground with its stern warnings about the risk of future climate shocks (Chapter 4). The conference report became required reading for the delegates at the United Nations' first major conference on the environment, held the following year. Heeding the scientists' recommendations, the conference set in motion a vigorous program of cooperative research on the environment, with climate research included. Meanwhile the GARP committee set up a series of large-scale "experiments" that coordinated a great variety of government and academic institutions. An outstanding example was a project conducted in 1974, the GARP Atlantic Tropical Experiment (GATE, an acronym containing an acronym!). That summer, 40 research ships and a dozen aircraft from 20 nations made measurements across a large swath of the tropical Atlantic Ocean, studying the flows of heat and moisture into the atmosphere.

That was the easy part. However complex the oceans and atmosphere might be, research on them followed well-defined tracks. But the more scientists looked at the climate system from an international perspective, the more they noticed that there were additional components, even more complicated and scarcely studied at all. They began to discover evidence that the forests of Africa, the tundra of Siberia, and other living ecosystems were somehow essential parts of the climate system. How did they fit in?

Little was known about connections between the planet's biomass and the atmosphere. The few people who looked into the question found that the amount of carbon in the atmosphere is only a fraction of the amount bound up in trees, peat bogs, soils, and other products of terrestrial life. These ecosystems and their stock of organic carbon seemed to have been fairly stable over millions of years. The likely cause of stability was a fact demonstrated by experiments in greenhouses and in the field: plants often grow more lushly in air that is "fertilized" with extra CO_2. Thus if more of the gas were added to the atmosphere, it should be rapidly taken up, made into wood and soil. This was one more version of the argument that the atmosphere was automatically stabilized, part of the indestructible Balance of Nature.

Most geological experts thought that even on a lifeless planet, the atmospheric balance would remain stable. It seemed reasonable to expect that chemical cycles would long ago have settled down into some kind of equilibrium among air, rocks, and seawater. Compared with those titanic kilometer-thick masses of minerals, it hardly seemed necessary to consider the thin scum of bacteria and so forth. Thus, for example, a 1966 Academy of Sciences study of climate change concentrated on cities and industry; the panel remarked that changes to the countryside, such as irrigation and deforestation, were "quite small and localized," and set that topic aside without study.[11]

As evidence mounted that global harm could be inflicted by such human products as chemical pesticides or dust, the traditional belief in the automatic stability of biological systems faltered. Concerns were redoubled by the African drought of the early 1970s. Was the Sahara desert expanding southward as part of a natural climate cycle that would soon reverse itself, or was something more ominous at work? For a century, African travelers and geographers had worried that overgrazing could cause such changes in the land that "man's stupidity" would create a "man-made desert."[12] In 1975, veteran climate scientist Jule Charney proposed a mechanism. Noting that satellite pictures showed widespread destruction of African vegetation from overgrazing, he pointed out that the barren clay reflected sunlight more than the grasses had. He figured this increase of albedo would make the surface cooler, and that might change the pattern of winds so as to bring less rain. Then more plants would die, and a self-sustaining feedback would push on to full desertification.

Charney was indulging in speculation, for computer models of the time were too crude to show what a regional change of albedo would actually do to the winds. It would be a few more years before models demonstrated that vegetation is indeed an important factor in a region's climate. But it didn't require detailed proof for scientists to grasp the truth of Charney's primary lesson. Human activity could change vegetation enough to affect the climate. The biosphere did not necessarily regulate the atmosphere smoothly, but could itself be a source of instability.

The public was meanwhile learning how slash-and-burn farming was eating its way through entire tropical forests, and that only a diminishing remnant remained of the great ancient forests of North America. Concern about these losses was rising, though for the sake of wildlife rather than climate. Meanwhile a few scientists pointed out that these forests were a significant player in global cycles of carbon and water. A forest evaporating moisture can be wetter than an ocean, in its effect on the air above it. But just what kind of changes would deforestation bring? The answers lay in an uncharted no-man's-land between the very different fields of meteorology and biology.

Only a few things could be measured with confidence. Statistics compiled by governments on the use of fossil fuels told how much CO_2 was going into the atmosphere from industrial production. And Keeling's measurements, pursued indefatigably decade after decade, showed how much of the gas remained in the air, shoving the curve higher year by year. The two numbers were unequal.

More than half of the gas from burning fossil fuels was missing. Where was the missing carbon going?

There were only two likely suspects. The carbon must wind up either in the oceans, or in biomass. In the early 1970s, Wally Broecker and others developed models for the movement of carbon in the oceans, including the carbon processed by living creatures. They calculated that the oceans were taking up much of the new CO_2, but not all of it. The residue must somehow be sinking into the biosphere. Perhaps trees and other plants were growing more lushly thanks to CO_2 fertilization?[13]

That was hard to check. Few solid studies of fertilization had been published by plant biologists—a type of specialist that had scarcely interacted with climate scientists. And as Keeling admitted in 1973, even with good data on past conditions, any calculation of the present or future fertilizer effect would be unreliable. Every gardener knows that giving a plant more fertilizer will promote growth only up to a certain level. Nobody knew where that level was if you gave more CO_2 to the world's various kinds of plants. "We are thus practically obliged to consider the rate of increase of biota as an unknown," Keeling warned.[14] Some rough calculations suggested that land plants might not be a sink at all. The decay of organic matter in soils was increased by deforestation and other human works, so the land biota could be a major net *source* of the gas.

The uncertainties became painfully obvious at a workshop held in Dahlem, Germany, in 1976. Bolin argued that human damage to forests and soils was releasing a very large net amount of CO_2. Since the level in the atmosphere was not rising all that fast, it followed that the oceans must be taking up the gas, much more effectively than geochemists like Broecker had calculated. George Woodwell, a botanist who studied ecosystems, went still farther with his own calculations. He argued that deforestation and agriculture were putting into the air as much CO_2 as the total from burning fossil fuel, or maybe even twice as much. His message was that our attack on forests should be halted, not just for the sake of preserving natural ecosystems, but also to preserve the climate.

Broecker and his colleagues thought Woodwell was making ridiculous extrapolations from scanty data. Defending their own calculations, the oceanographers and geochemists insisted that the oceans could not possibly be taking up so much carbon. The scientists debated one another vigorously at the Dahlem Conference. The arguments spilled over into social questions, all the issues of environmentalism and government intervention that were raised by deforestation and desertification. People's beliefs about the sources of CO_2 were becoming connected to their beliefs about what actions (if any) governments should take. Woodwell insisted that tropical deforestation and other assaults on the biosphere were "a major threat to the present world order."[15] He publicly called for a halt to burning forests, as well as aggressive reforestation to soak up excess carbon.

Such discussions were no longer restricted to scientists and the mid-level government officials they dealt with. Saving the forests had become a popular idea in the growing environmental movement (in which Woodwell played a prominent role as organizer). The forestry and fossil-fuel industries took note, recognizing that worries about greenhouse gases might lead to government regulation. They were joined by political conservatives, who tended to lump together all claims about impending ecological dooms as left-wing propaganda.

When environmentalist ideals had first stirred, around the time of Theodore Roosevelt, they had been scattered across the entire political spectrum. A traditional conservative, let us say a Republican birdwatcher, could be far more concerned about "conservation" than a Democratic steelworker (in more recent times, at the far end of the traditional Left, Communist nations were the planet's most egregious polluters). But during the 1960s, as the new Left rose to prominence, it became permanently associated with environmentalism. Perhaps that was inevitable. Many environmental problems, like smog, seemed impossible to solve without government intervention. Such interventions were anathema to the new Right that began to ascend in the 1970s.

By the mid-1970s, conservative economic and ideological interests had joined forces to combat what they saw as mindless eco-radicalism. Establishing conservative think tanks and media outlets, they propagated sophisticated intellectual arguments and expert public-relations campaigns against government regulation for any purpose whatsoever. On global warming, it was naturally the fossil-fuel industries that took the lead. Backed up by some scientists, industry groups developed arguments ranging from elaborate studies to punchy advertisements, aiming to persuade the public that there was nothing to worry about.

As environmental and industrial groups hurled uncompromising claims back and forth across a widening political gulf, most scientists found it hard to get a hearing for more ambiguous views. Journalists in search of a gripping story tended to present every scientific question as if it were a head-on battle between two equal and diametrically opposite sides. Yet most scientists saw themselves as just a bunch of people with various degrees of uncertainty, groping about in a fog.

Faced with the controversy over carbon emitted by deforestation, researchers tried to resolve the problem scientifically. In meetings and publications the experts wrangled, sometimes vehemently but always courteously. As occasionally happens in scientific debates, opinions tended to divide along disciplinary lines: oceanographers plus geochemists versus biologists. The physical scientists like Broecker pointed out that their models of the oceans could be reliably calibrated with data on how the waters took up radioactive materials (fallout from nuclear weapon tests was especially useful). Woodwell's biology was manifestly trickier. His opponents argued that nobody really knew what was happening to the plants of the Amazon and Siberia. When he invoked field studies carried out in this or that patch of trees, his opponents brought up more ambiguous studies, or just said that studies of a few hectares here and there could scarcely be extrapolated to all the world's forests.

The key data finally came from measurements of radioactive carbon (using the fact that newly created isotopes cycled through the atmosphere and plants, whereas fossil-fuel emissions had long since lost their radioactivity). The ocean models turned out to be roughly correct. The gas emitted from decaying or burned plants was more or less balanced by the amount taken up by other plants. Perhaps deforestation was compensated by more vigorous growth resulting from fertilization by the increased CO_2 in the atmosphere. Woodwell denied this, but other scientists gradually concluded that his claims were exaggerated. Eventually he had to concede that deforestation was not adding as much CO_2 to the atmosphere as he had thought. Much remained unexplained, and nobody was sure just how to balance the global carbon budget.

An important lesson remained. As a team headed by Broecker wrote in 1979, Woodwell's claims that destruction of plants released huge amounts of CO_2 had been a "shock to those of us engaged in global carbon budgeting." The intense reexamination triggered by the claim had called attention to "the potential of the biosphere."[16] From the late 1970s onward, it was clear that nobody could predict the future of global climate with much precision until they could say how the planet's living systems affected the level of CO_2. Of course, to answer that you would have to know how the biosphere itself would change if the atmosphere changed. And to answer that you would have to know how the atmosphere would respond to changes in the oceans, and ice sheets, and more.

The great trick of science is that you don't have to understand everything at once. Scientists are not like the people who have to make decisions in, say, business or politics. Scientists can pare down a system into something so simple that they have a chance to understand it. That is, if they dedicate their lives to the task. That was the approach taken by many, among them Syukuro Manabe.

"Suki" Manabe was one of a group of young men who graduated from Tokyo University in the difficult years just after the Second World War and looked for a career in meteorology. Ambitious and independent-minded, they had few opportunities for advancement in Japan and wound up making their careers in the United States. In 1958 Manabe was invited to join the computer modeling group founded by John von Neumann. After the group had achieved its breakthrough in 1955 with a model that produced realistic-looking regional weather (Chapter 3), von Neumann had drummed up government funding for a project with an ambitious goal. His team would construct a general circulation model of the entire three-dimensional global atmosphere, deriving climate directly from the basic physics equations for fluids and energy. The effort got under way in 1948 at the U.S. Weather Bureau in Washington, D.C., under the direction of Joseph Smagorinsky.

Smagorinsky's best idea was to recruit Manabe and work with him on the proposed climate model. Starting with winds, rain, snow, and sunlight, they added the greenhouse effect of water vapor and CO_2; they put in the way moisture and heat were exchanged between air and the surfaces of ocean, land, and ice, and much more. Manabe spent many hours in the library researching such esoteric topics as the way various types of soil absorbed water. The equations too needed attention: the system had to be efficient to calculate, not prone to veer off into wildly unrealistic numbers.

By 1965 Manabe and Smagorinsky had a three-dimensional model that solved the basic equations for an atmosphere divided into nine levels. It was highly simplified, with no geography—everything was averaged over zones of latitude, and land and ocean surfaces were blended into a sort of swamp that exchanged moisture with the air but could not take up heat. Nevertheless the way the model moved water vapor around the planet looked gratifyingly realistic. The printouts showed a stratosphere, a zone of rising air near the equator (creating the doldrums that becalmed mariners), a subtropical band of deserts, and so forth. Many of the details came out wrong, however.

As it became apparent that such modeling could bring useful results, more groups joined the effort. A post-doctoral student might take a job at a new institution, bringing along his former group's computer code, and assemble a new team to modify and improve it. Others built their models from scratch. Central to their progress was the headlong advance of electronic computers: from the mid-1950s to the mid-1970s, the power available to modelers increased by a factor of several thousand. Through the 1960s and 1970s, important general circulation model (GCM) groups appeared at institutions from New York to Australia.

Particularly influential was a group at the University of California, Los Angeles, where Yale Mintz had recruited another young Tokyo University graduate, Akio Arakawa, to help with the mathematics. In 1965, Mintz and Arakawa produced a model which, like Smagorinsky and Manabe's, bore some resemblance to the real world. Another important effort was started in 1964 at the National Center for Atmospheric Research in Boulder, Colorado. The leaders were Warren Washington and yet another Tokyo graduate, Akira Kasahara. Funded by the National Science Foundation and run by a consortium of universities, NCAR became one of the world's chief centers for climate modeling. But the premier model was Manabe's, in Smagorinsky's Weather Bureau unit (it was eventually renamed the Geophysical Fluid Dynamics Laboratory and moved to Princeton).

Although modeling had become a large combined effort, it was still not able to force through the thicket of problems. The best computers of the day, using up weeks of costly running time, could calculate only a very crude simulation of a typical year of global climate. The workers figured they would need a hundred times more computer power to do the job right. They would get that within another couple of decades. Yet even if the computers had been a million times faster, the simulations would still have been unreliable. For the modelers were still facing that famous limitation of computers, "garbage in, garbage out."

The calculations depended crucially, for example, on what sort of clouds would grow under certain conditions. The fastest computers even today are far from able to calculate the details of every cloud on the planet. They must settle for figuring the average behavior of clouds in each cell of a grid, where the cell is hundreds of kilometers wide. Modelers had to develop "parameterizations," working up a set of numbers (parameters) to represent the net effects of all the clouds in a cell under given conditions. To get these numbers, the modelers had only some basic equations, a scattering of unreliable data, and plenty of guesswork.

If that obstacle had somehow gone away, another remained. To diagnose the failings that kept models from being more realistic, scientists needed an abundance of data—the actual profiles of wind, heat, moisture, and so on at every level of the atmosphere and all around the globe. The data in hand through the 1960s were shamefully sparse. Smagorinsky put the matter succinctly in 1969: "We are now getting to the point where the dispersion of simulation results is comparable to the uncertainty of establishing the actual atmospheric structure."[17]

Help came from the drive to improve short-term weather predictions. By around 1970 computer models were giving better results than the old rule-of-thumb forecasters, predicting weather as far as three days ahead. That meant cash to farmers and other businesses, and the work attracted ample funding. The forecasting models required data on conditions at every level of the atmosphere at thousands of points around the world. Such observations were now being provided by the balloons and sounding rockets of the international World Weather Watch. Still better help came from outer space.

Use of satellites for weather "reconnaissance" had been proposed in a secret report as early as 1950, and the first public satellite to monitor global weather had been built under a Department of Defense program and launched in 1960. Through the following decades, this program continued to build and operate secret meteorological satellites, using the exquisite and highly classified technologies developed for spy satellites. These technologies were gradually transferred to an open civilian program. When computer modelers reached the point where they could not progress without much better data on the actual atmosphere, the answer was Nimbus-3, launched in 1969. The satellite's infrared detectors could measure the temperature of the atmosphere comprehensively at various levels, night and day, over oceans, deserts, and tundra. Once again science was benefiting from money spent for practical military and civilian purposes.

The work settled into steady, tenacious improvement of existing techniques. Modelers put in ever more factors, filling in the most gaping holes, and developed ever more efficient ways to use their rapidly improving computers. Encouragement came from a 1972 model by Mintz and Arakawa, which managed to simulate roughly the huge changes as the sunlight shifted from season to season. During the next few years Manabe and collaborators published a model that produced entirely plausible seasonal variations. That was a convincing test of the models' validity. It was almost as if a single model worked for two quite different planets—the planets Summer and Winter.

NASA's Goddard Institute for Space Studies in New York City took a different approach. The group had been developing a weather model as a practical application of its mission to study the atmospheres of planets. James (Jim) Hansen assembled a team to reshape their equations into a climate model. By simplifying some features while adding depth to others, they managed to get a quite realistic-looking simulation that ran an order of magnitude faster than rival general circulation models. That permitted the group to experiment with multiple runs, varying one factor or another to see what changed. In such studies, the global climate was beginning to look and feel to researchers like a comprehensible physical system, akin to the systems of glassware and chemicals that experimental scientists manipulated on their laboratory benches.

Sophisticated computer models were steadily displacing the traditional simplistic hand-waving models. In the digital models, it was clear from the outset that climate was the outcome of a staggeringly intricate complex of interactions and feedbacks among many global forces. No easy explanation was offered for even such a simple feature of the atmosphere as the windless semitropical zones that sailors called the doldrums. *In principle* the shape of the general circulation could only be comprehended indirectly in the working-through of a million calculations.

In their first decade or so of work, the modelers had a hard enough time just trying to understand a typical year's average weather. But in the mid-1960s, a few of them had begun to take an interest in the question of what could cause changes in climate. They saw Keeling's curve of rising CO_2, and they saw Fritz Möller's discovery that simple models built out of a few equations showed disturbing instabilities. When Möller visited Manabe and explained his grotesque results, Manabe decided to look into how the climate system might actually change.

Manabe and his collaborators were already building a model that included, among many other things, the way air and moisture conveyed heat from the Earth's surface into the upper atmosphere. That was a big step beyond trying to calculate surface temperatures, as Möller and others since Arrhenius had attempted, just by considering the energy balance at the surface. To get a really sound answer, the atmosphere had to be studied as a tightly interacting system from top to bottom. In such a model, when the surface warmed, updrafts would carry heat by convection into the upper atmosphere—so the surface temperature would not run away as in Möller's model. The required computations were so extensive, however, that Manabe had to strip the model down to a single one-dimensional column, representing a slice of atmosphere averaged over the entire globe or over a single band of latitude.

In 1967, Manabe's group used this model to test what would happen if the level of CO_2 in the atmosphere changed. Their target was something that would eventually become a central preoccupation of modelers: climate "sensitivity," that is, how much average global temperature would be altered by a given change in one variable (the Sun's output of energy, say, or the CO_2 level). They would run a model with one value of the variable (say, of CO_2), run it again with a different value, and compare the answers. Ever since Arrhenius, researchers had pursued this question with highly simplified calculations. They used as a benchmark the difference if the CO_2 level was doubled. After all, Keeling's curve showed that the level would probably double sometime in the twenty-first century. The number Manabe's group came up with for doubled CO_2 was a rise of global temperature of roughly 2°C (around 3 to 4°F). This was the first time a greenhouse-effect warming calculation included enough of the essential factors to seem reasonable to many experts. As Broecker for one recalled, it was the 1967 paper "that convinced me that this was a thing to worry about."[18]

This model of a one-dimensional column of air was a far cry from a full three-dimensional general circulation model. Using the column as a basic building block, Manabe and a collaborator, Richard Wetherald, constructed such a GCM in the early 1970s. It was still highly simplified. In place of actual land and ocean geography, they pictured a planet that was half land and half swamp. But overall, this mock planet had a climate system that looked pretty much like the Earth's. In particular, the reflection of sunlight from the model planet at each latitude agreed pretty well with the actual numbers for the Earth, as measured by the new Nimbus-3 weather satellite. For doubled CO_2 the computer predicted an average warming of around 3.5°C. When they published this in 1975, Manabe and Wetherald warned that the result should not be taken too seriously; the model was still scarcely like a real planet. A better prediction of climate change would have to wait on general improvements.

Above all, there was the vexing problem of clouds. As the planet got warmer, the amount of cloudiness would probably change, but change how? There was no trustworthy way to figure that out. And what would a change in clouds mean for climate? Scientists were beginning to realize that clouds could either cool a region (by reflecting sunlight) or warm it (by trapping heat radiation from below). It depended on the type of clouds, and how high they floated in the atmosphere. Worse still, it was becoming clear that the way clouds formed could be strongly affected by shifts in the haze of dust and chemical particles floating in the atmosphere. Little was known about how such aerosols helped or hindered the formation of different types of clouds.

The uncertainty was unacceptable, for people had begun to demand much more than a crude reproduction of the present climate. After the weather disasters and energy crisis of the early 1970s put greenhouse warming on the political agenda (for people who paid attention to technical issues), it became a matter of public debate whether the computer models were correct in their predictions of global warming. Must we halt deforestation, and turn away from fossil fuels? News reports featured disagreements among prominent scientists, especially over whether warming or cooling was likely. "Meteorologists still hold out global modeling as the best hope for achieving climate prediction," a senior scientist observed in 1977. "However, optimism has been replaced by a sober realization that the problem is enormously complex."[19]

One of the most troubling complexities was explained by the prominent meteorologist William W. Kellogg. In 1975 he pointed out that industrial aerosols, as well as the soot from burning debris where forests were cleared, strongly absorbed sunlight—after all, smog and smoke are visibly dark—and retained heat. Hence, he argued, the main effect of human aerosols would be regional warming; there was thus no need to worry about cooling from pollution. Bryson and his co-workers continued to insist that smoke and haze had a powerful cooling effect—after all, they visibly dimmed sunlight. The debate was hard to resolve. Nobody could calculate from basic physics principles, under various circumstances, whether a haze would bring warming or cooling.

Nevertheless, in the choice between warming and cooling, scientific opinion was beginning to come down on one side. Perhaps in the natural course of events the Earth would gradually slide into an ice age—but the course of events was no longer natural. More and more scientists felt that the greenhouse effect was the main thing to worry about. Confidence could never come from a single line of attack, but various kinds of work all pointed in the same direction.

Where calculations failed, more roundabout methods might offer answers. For example, Stephen Schneider and a collaborator studied the effects of dust by matching the temperature record of the past 1000 years with volcanic eruptions. Their simplified model predicted that CO_2 warming would dominate surface temperature patterns soon after 1980. As Kellogg had also pointed out, rains washed aerosols out of the lower atmosphere in a matter of weeks. And many nations were vigorously working to lower air pollution. So no matter whether aerosols warmed the Earth or cooled it, the greenhouse effect from increased CO_2—a gas which would linger in the atmosphere for centuries—must necessarily dominate in the end.

In 1977 the National Academy of Sciences weighed in with a major study by a panel of experts. Their consensus was that cooling was not likely over the long term. Rather, temperatures might rise to nearly catastrophic levels during the next century or two. The panel's report, announced at a press conference during the hottest July the nation had experienced since the Dust Bowl years of the 1930s, was widely noted in the press. Science journalists, by now closely attuned to the views of climate scientists, promptly reflected the shift of opinion. When, in 1976, *Business Week* had explained both sides of the debate, it reported that "the dominant school maintains that the world is becoming cooler." Just one year later, the magazine declared that CO_2 "may be the world's biggest environmental problem, threatening to raise the world's temperature" with horrendous long-term consequences.[20]

To get a more authoritative answer, the President's Science Adviser asked the National Academy of Sciences to study whether GCMs were trustworthy. The Academy appointed a panel, chaired by Charney and including other respected experts who had been distant from the recent climate debates. Their conclusion was unequivocal: when it came down to the main issue, the models were telling the truth. To make their conclusion more concrete, the panel decided to announce a specific range of numbers. Splitting the difference between Hansen's GCM, which predicted a 4°C rise for doubled CO_2, and Manabe's latest figure of around 2°C, the Charney panel declared they had rather high confidence that in the next century the Earth would warm up by about three degrees plus or minus fifty percent, that is, 1.5–4.5°C (2.7–8°F). They concluded dryly, "We have tried but have been unable to find any overlooked or underestimated physical effects" that could reduce the warming. "Gloomsday Predictions Have No Fault" was how *Science* magazine summarized the report.[21]

Many climate experts felt less confident. The computer models showed toy planets scarcely resembling our Earth. They were flat geometrical constructions without mountains or other real geography, with stagnant swamps in place of oceans, and clouds made up from guesswork. There was no way to prove that some of the many pieces the models left out were not crucial. The long-term trend toward the next ice age remained plausible in the absence of human intervention, and human intervention might have all sorts of unexpected effects. Journalists could track down experts who would give strong and definite statement about such questions, but most scientists were willing to live out their careers in a state of uncertainty. They would watch as studies accumulated, seeing which ones reinforced or contradicted one another. On a subject as complex as climate change, no single finding would radically change their opinions. One year a given expert might feel it was 60 percent likely that warming would come; after a few years that might shift down to 50 percent or up to 70 percent.

The opinions of virtually all important climate experts came up against one another at a World Climate Conference that was held in Geneva in 1979. Well in advance, the conference organizers commissioned a set of review papers inspecting the state of climate science, and these were circulated, discussed, and revised. Then more than 300 experts from over 50 countries assembled to examine the review papers and recommend conclusions. The experts' views about what might happen to the climate spanned a broad spectrum, yet they managed to reach a consensus. In their concluding statement, the scientists at the conference recognized a "clear possibility" that an increase of CO_2 "may result in significant and possibly major long-term changes of the global-scale climate." This cautious indication of an eventual "possibility" was hardly news, and it caught little public or political attention.

As the 1980s began, the possibility of global warming had become prominent enough to be included for the first time in public opinion polls. A 1981 survey found that over a third of American adults claimed they had heard or read about the greenhouse effect. That meant the news had spread beyond the small minority who regularly followed scientific issues. When pollsters explicitly asked people what they thought of "increased carbon dioxide in the atmosphere leading to changes in weather patterns," nearly two-thirds replied that the problem was "somewhat serious" or "very serious."[22]

Most of these citizens would never have brought up the subject by themselves. Only a small fraction understood that the risk of climate change was mainly due to carbon dioxide from fossil fuels. They blamed smog or other chemical pollution, nuclear tests, even spaceship launches. And those who worried most about the environment were seldom concerned with global affairs, directing their dismay at the oil spill or chemical wastes that endangered a particular neighborhood. Many people now suspected that global warming was something they ought to be concerned about, but among the world's many problems it did not loom large.

CHAPTER SIX

THE ERRATIC BEAST

Ed Lorenz thought the climate might move in any direction, with little warning. His work in meteorology had helped lay the foundation of the newly fashionable chaos theory, and he continued to take the lead in studying how tiny initial variations could tilt a complex system this way or that. At a 1979 meeting he asked a famous question: "Does the flap of a butterfly's wings in Brazil set off a tornado in Texas?" His answer—perhaps it could—became part of the common understanding of educated people.[1]

Climate change had once seemed a simple enough concept, a gradual evolution responding to a direct push, whether your favorite theory said the push was a change in sunlight or volcanic haze or the level of CO_2. Decade after decade, scientists had discovered complications. The uncertain possibilities of industrial pollution, surging ice sheets, and deforestation were bad enough, but in the late 1970s and 1980s even more factors turned up. The climate began to look less like a simple mechanical system than like a confused beast which a dozen different forces were prodding in different directions.

Lorenz said the outcome might be unpredictable in principle. He and others argued that the warming and cooling trends of the past century might not be evidence of responses to aerosols or a greenhouse effect or anything else in particular. It might be only the beast erratically lunging this way and that, in obedience to its incalculably complex internal reaction to the various external pressures. Most scientists agreed that climate has features of a chaotic system, but they did not think it was wholly random. It might well be impossible in principle to predict that a tornado would hit a particular town in Texas on a particular day (not because of one guilty butterfly, of course, but as the net result of countless tiny initial influences). Yet tornado seasons came on schedule. That type of consistency showed up in the computer simulations constructed in the 1980s. Start a variety of GCM runs with different initial conditions, and they would show random variations in the weather patterns computed for one or another region and season. But the runs would converge when it came to average annual global temperature. And every model ended with some sort of warming over the next century.

This was not good enough for the critics, who noted many points where all the models shared uncertain assumptions and flimsy data. The modelers admitted that they still had a long way to go. Their GCMs, debatable and arcane, could inspire little confidence in policy-makers and the public. People thinking about climate change wanted a more straightforward indicator—like the weather outside their windows. It was scarcely possible to get the public, or even most scientists, to take global warming seriously, if the average temperature of the planet was dropping.

But was it? In 1975 two New Zealand scientists reported that while the Northern Hemisphere had been cooling over the past thirty years, their own region, and probably other parts of the Southern Hemisphere, had been warming. There were too few weather stations in the vast unvisited southern oceans to be certain, but other studies tended to confirm it. The cooling since around 1940 had been observed mainly in northern latitudes. Perhaps the greenhouse warming was counteracted there by cooling from industrial haze? After all, the Northern Hemisphere was home to most of the world's industry. It was also home to most of the world's population, and as usual, people had been most impressed by the weather where they lived.

Scientists and government policy-makers needed to know for sure what had been happening to the weather. Thousands of stations around the world had churned out daily numbers, but the numbers conformed to no single standard: they made an almost indecipherable muddle. Around 1980, two groups undertook to work through the numbers in all their messy historical and technical details, rejecting sets of uncertain data and tidying up the rest.

First to weigh in was Jim Hansen's group in New York. They reported that "the common misconception that the world is cooling is based on Northern Hemisphere experience to 1970." Just around the time that meteorologists had noticed the cooling trend, it had apparently reversed. From a low point in the mid-1960s, by 1980 the world as a whole had warmed some 0.2°C.[2] The temporary northern cooling from the 1940s through the 1960s had been bad luck for climate science. By feeding skepticism about the greenhouse effect, and by provoking some scientists and many journalists to speculate publicly about the coming of a new ice age, the cool spell gave the field a reputation for fecklessness that it would not soon live down.

Any greenhouse warming would be masked not only by random natural variations and industrial pollution, but also by some fundamental planetary physics. If anything added heat to the atmosphere, much of it would be absorbed into the upper layer of the oceans. While that was warming up, perception of the problem would be delayed. The Charney panel of experts had explained the effect in 1979: "We may not be given a warning until the CO_2 loading is such that an appreciable climate change is inevitable."[3] Uptake of heat by the oceans could only delay atmospheric warming by a few decades. In 1981, Hansen's group boldly predicted that, considering how fast

CO_2 was accumulating, "carbon dioxide warming should emerge from the noise level of natural climatic variability" by the end of the century. Other scientists, using different calculations, agreed. The discovery of global warming—that is, plain evidence that the greenhouse effect really operated as predicted—would come sometime around the year 2000.[4]

The second group analyzing global temperatures was the British government's Climatic Research Unit at the University of East Anglia, led by Tom M. L. Wigley and P. D. Jones. In 1986 they produced the first entirely solid and comprehensive global analysis of average surface temperatures. The warmest three years in their 134-year record had all occurred in the 1980s. Convincing confirmation came from Hansen and a collaborator, who analyzed old records using quite different methods from the British and came up with substantially the same results. It was true: an unprecedented warming was under way.

In such publications, the few pages of text and graphs were the visible tip of a prodigious unseen volume of work. Many thousands of people in many countries had spent much of their lives measuring the weather, while thousands more had devoted themselves to organizing and administering the programs, improving the instruments, standardizing the data, and maintaining the records in archives. In geophysics not much came easily. One simple sentence (like "last year was the warmest year on record") might be the distillation of the labors of a multigenerational global community. And it still had to be interpreted.

Most experts saw no solid proof that continued warming lay in the future. After all, reliable records covered little more than a century, and they showed large fluctuations (especially the 1940–1970 dip). Couldn't the current trend be just another temporary wobble? Schneider, one of the scientists least shy about warning of climate dangers, acknowledged that "a greenhouse signal cannot yet be said

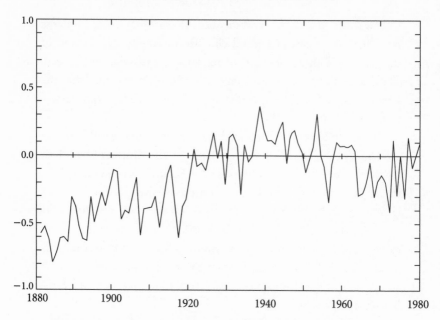

Figure 2. THE ERRATIC RISE OF TEMPERATURE.

The average surface temperature of the Northern Hemisphere (shown as differences from the 1946–1960 mean in °C). This 1982 graph by a British group shows the prominent rise to the 1940s, confusing decline through the 1960s, and vague beginnings of the subsequent rise (which soared much higher in the 1990s, see Figure 3 in Chapter 8). (P. D. Jones et al., *Monthly Weather Review* 110, 1982, p. 67, reproduced by permission of American Meteorological Society.)

to be unambiguously detected in the record." Like Hansen, he expected that the signal would only emerge clearly around the end of the century.[5]

The great, unyielding obstacle to a reliable understanding of greenhouse warming was the natural variation of climate. The way temperatures had soared in some decades, then dipped in others, showed that factors besides the greenhouse effect must be at work. When modelers like Hansen added a factor for volcanic eruptions to the rise of CO_2, they found this could explain a good part of the variation—but not all. What else was pushing the atmosphere around?

There were hints that part of the variation followed a regular cycle. For example, Dansgaard's cores of ancient ice drilled from deep in the Greenland ice sheet showed a cycle about 80 years long. The Danish team supposed the Sun was responsible. Just such a cycle had already been reported by scientists who had analyzed small variations in the number of sunspots. Among those impressed by these findings was Broecker. He figured that a downtrend in the solar cycle, even more than volcanoes and industrial haze, had been temporarily canceling out greenhouse-effect warming. In 1975 he published an influential article suggesting that once the solar cycle turned, while CO_2 continued to accumulate, the world might be poised on the brink of a serious rise of temperature. "Complacency may not be warranted," he said. "We may be in for a climatic surprise."[6]

Later studies failed to find Dansgaard's cycles globally: insofar as they existed at all, they seemed to be caused by something that happened in the North Atlantic Ocean, not on the Sun. It was just another case of supposed global weather cycles that faded away as more data came in. It was also one of several cases where Broecker's scientific instincts outstripped his evidence. For there were indeed reasons to suspect that changes in the Sun affected climate.

Back in the early 1960s one of the new radiocarbon experts, Minze Stuiver, had moved in the right direction. In collaboration with another radiocarbon expert, Hans Suess, he had showed that the amount of radiocarbon in ancient tree rings varied from century to century. Many scientists were upset, for the erratic variations undermined the technique of radiocarbon dating. But what looks like unwelcome noise to one specialist may contain information for another. Stuiver noted that radiocarbon is generated in the atmosphere by cosmic rays from the far reaches of the universe, and he pointed out that the magnetic field of the Sun impedes the flux of cosmic rays reaching the Earth. Perhaps the radiocarbon record said something about changes on the Sun?

In 1965 Suess tried correlating the new data with weather records, in the hope that radiocarbon variations "may supply conclusive evidence regarding the causes for the great ice ages."[7] He focused on a bitter cold spell reported in European weather records of the sixteenth and seventeenth centuries: the Little Ice Age, when crops had failed repeatedly and the Thames at London froze solid in winter. That had been a period of relatively high radiocarbon. Casting a sharp eye on historical sunspot data, Suess noticed that around the time of the Little Ice Age, few sunspots had been recorded. Fewer sunspots indicated a lower level of solar magnetism, which meant more cosmic rays could get through, which created more radiocarbon. In short, Suess suggested, more radiocarbon pointed to changes on the Sun which somehow connected with cold winters.

Some found the connections plausible, but to most scientists the speculation sounded like just one more of the countless sunspot correlations that had been announced only to be rejected sooner or later. Even if the evidence had been stronger, it would have met with deep skepticism, for scientists cannot well fit data into their thinking unless theory meanwhile prepares a place. The variations of sunspots and cosmic rays were negligible compared with the Sun's total output of energy. How could such trivial variations possibly have a noticeable effect on climate?

In 1975 the respected meteorologist Robert Dickinson took on the task of reviewing the American Meteorological Society's official statement about solar influences on weather. He concluded that such influences were unlikely, for there was no reasonable mechanism in sight—except, maybe, one. Perhaps the electrical charges that cosmic rays brought into the atmosphere somehow affected the way cloud droplets condensed on dust particles. Dickinson hastened to point out that this was pure speculation. Scientists knew very little about how clouds formed, and would need to do much more research "to be able to verify or (as seems more likely) to disprove these ideas."[8] For all his frank skepticism, Dickinson had left the door open a crack. One way or another, it was now at least scientifically conceivable that changes in sunspots could have something to do with changes in climate.

In 1976 a solar physicist tied all the threads together in a paper that soon became famous. John (Jack) Eddy was one of several solar physicists in Boulder. Despite the variety of climate experts in that city, Eddy was ignorant of the radiocarbon research—an example of the poor communication between fields that always impeded climate studies. Eddy had had scant success in the usual sort of solar physics research, and in 1973 he lost his job as a researcher, finding only temporary work writing a history of NASA's Skylab. In his spare time he pored over old books. He had decided to review historical naked-eye sunspot records with the aim of definitively confirming the long-standing belief that the sunspot cycle was stable over the centuries.

But the records indicated that the Sun was by no means constant. This discovery, like many in science, was not entirely new. An almost complete absence of sunspot observations during the Little Ice Age had been noticed not only by Suess but by several others before him. In particular, back in 1890 a British astronomer, E. Walter Maunder, had drawn attention to the evidence and a possible climate connection. Other scientists had thought this was just another case of dubious numbers at the edge of detectability, and Maunder's publications sank into obscurity. A scientific finding cannot flourish in isolation but needs support from other findings.

"As a solar astronomer I felt certain that it could never have happened," Eddy later recalled.[9] But hard historical work gradually persuaded him that the early modern solar observers were reliable—the absence of sunspot evidence really was evidence of an absence. Other scientists were skeptical, but as Eddy pushed his arguments, he learned of the radiocarbon evidence. Stuiver and others had meanwhile confirmed the connection between solar activity and the radiocarbon in tree rings and other fossil sources. When Eddy presented his full results in 1976, correlating sunspots, radiocarbon, and temperatures, many found the evidence convincing. He warned that in our own times, "when we have observed the Sun most intensively, its behavior may have been unusually regular and benign."[10]

The next step was to look for confirmation. Schneider, Hansen, and others found that they could indeed get a decent match to past temperature trends if they took into account not only the sporadic cooling due to dust from volcanic eruptions, but also the solar variability indicated by sunspots and radiocarbon. Adjusting the strength of the presumed solar influence to match the historical temperature curve was guesswork, dangerously close to fudging. But sometimes a scientist must "march with both feet in the air," assuming a couple of things at once in order to see whether it all eventually works out. The results looked good enough to encourage further studies. Some of these failed to find any correlations. As a reviewer commented in 1985, "this is a controversial topic," and the connection between solar variations and climate change remained "an intriguing but unproven possibility."[11]

Solar physics was not the only complication added to the study of the atmosphere in the 1970s. Another, quite different set of scientific specialties added an entirely new type of question. The history of climate science is full of unexpected linkages, but perhaps none so odd and tenuous as the events that drew attention to trace chemicals. The inquiry began with the concern over how pollution from supersonic transport airplanes might alter the stratosphere (Chapter 4). In 1973, Mario Molina and Sherwood Roland thought it might be interesting to take a look at what other chemical emissions might do. They were astonished to find that the minor industrial gases known as CFCs (chlorofluorocarbons) could have serious effects.

Experts had thought that CFCs were environmentally sound. They were produced in relatively small quantities. And they were extremely stable, never reacting with animals and plants. It turned out that the stability itself made CFCs a hazard. They would linger in the air for centuries and eventually, Molina and Rowland realized, some would drift up to the stratosphere. There ultraviolet rays activated them and they became catalysts in a process that destroyed ozone. The high, thin layer of ozone blocks the Sun's ultraviolet rays, so removing this layer would result in an increase of skin cancers, and probably still worse dangers to people, plants, and animals.

CFCs were the propellents in aerosol sprays: every day millions of people were adding to the global harm as they used cans of deodorant or paint. Science journalists alerted the public, and environmentalists jumped on the issue. Industries fought back with public relations campaigns that indignantly denied there was any risk whatsoever. Unconvinced, citizens bombarded government representatives with letters, and the United States Congress responded in 1977 by banning the chemicals from aerosol spray cans. The issue had no visible connection with climate. But it sent a stinging message about how fragile the atmosphere was, how easily harmed by human pollution. And it showed that technical scientific findings about a future atmospheric risk could arouse the public enough to sway legislation and strike at major industries.

Rowland and Molina's earlier work on CFCs provoked V. Ram Ramanathan of NASA to take a closer look at these unusual molecules. In 1975 he reported that CFCs absorb infrared radiation prodigiously—they were greenhouse gases. A simple calculation suggested that CFCs, at the concentrations they would reach by the year 2000, all by themselves might raise global temperature by 1°C. Other scientists followed up with a calculation on other gases that had previously been little considered: methane (CH_4, the main component of natural gas) and nitrates (such as N_2O, emitted, for example, when fertilizer is spread about). If the level of these gases in the atmosphere doubled, it would raise the temperature another 1°C. In 1985 a team led by Ramanathan looked at some 30 trace gases that absorbed infrared radiation. These additional greenhouse gases put together, the team estimated, could bring as much global warming as CO_2 itself.

These gases had been overlooked because the amounts of them in the atmosphere were minuscule compared with CO_2. But there was already so much CO_2 in the air that the spectral bands where it absorbed radiation were mostly opaque already. You had to add a lot more CO_2 to make a serious difference. A few moments' thought would have told any physicist that it was otherwise for trace gases. For those, each additional wisp would help obscure a "window," a region of the spectrum that until then had let radiation through unhindered. But the simple is not always obvious unless someone points it out. Understanding took a while to spread. Well into the 1980s, the public, government officials, and even most scientists thought "global warming" was essentially synonymous with "increasing CO_2." Meanwhile many thousands of tons of other greenhouse gases were pouring into the atmosphere.

Some scientists did recognize the importance of methane and looked into its role in global carbon cycles. Methane is the swamp gas that bubbles out of bogs; it comes mostly from the bacteria that thrive everywhere, from garden soil to the guts of termites. These natural emissions were much greater than the amount of methane that escaped as humans extracted and burned natural gas. But that did not mean human effects were negligible. Humanity was imposing its will on much of the world's fertile surface, transforming the entire global biosphere. Specialists in obscure fields of research turned up a variety of methane sources that were rapidly increasing. The gas was emitted in geophysically significant quantities by the bacteria found, for example, in the mud of rice paddies, and in the stomachs of the proliferating herds of cows. And what about accelerated emissions from the soil bacteria, and even the termites, that flourished under deforestation and the advance of agriculture?

The effects turned out to be huge. In 1981 a group reported that methane in the atmosphere was increasing with astounding speed. A study of air trapped in bubbles in cores drilled from the Greenland icecap confirmed that a rise of methane had begun a few centuries ago, and wildly accelerated in recent decades. Through painstaking collection of air samples, the recent rise was measured accurately by 1988. The methane level had increased 11 percent in the past decade alone. And each molecule of the gas had a greenhouse effect about twenty times that of a molecule of CO_2.

This raised alarming possibilities for feedbacks. A huge reservoir of carbon is frozen in the deep permafrost layers of peat that underlie northern tundras. As arctic regions warmed up, would the endless expanses of sodden tundra emit enough methane to accelerate global warming? Even more ominous were the enormous quantities of carbon locked in the strange "clathrates" (methane hydrates), ice-like substances found in the muck of seabeds around the world. The clathrates are kept solid only by the pressure and cold of the overlying water. In the early 1980s it was pointed out that if a slight warming penetrated the sediments, the clathrates might melt and release colossal bursts of methane and CO_2 into the atmosphere—which would bring still more warming.

Of course there would also be feedbacks from normal biological production, as global warming altered the amounts of gases emitted or taken up by forests, grasslands, ocean plankton, and so forth. The only way to learn the outcome of such a complex system would be through computer modeling. A great deal had to be learned, however, before anyone could write equations that would accurately represent all the important effects.

To observe the climate working and changing as a single complex system, there was nothing so useful as the problem that had opened up the question in the first place: the ice ages. Understanding those mighty swings would take us a long way toward understanding the climate system. The prospects were good, for a crisp synopsis of past changes was emerging from continually improved studies of seabed clay and glacial ice. Technologies for working in the open ocean had been developed extensively for commercial purposes such as oil prospecting. In the 1970s, these technologies were put to use in a Deep Sea Drilling Project. This series of cruises, funded by the U.S. National Science Foundation, pulled up long cores from ocean floors around the world. The sequence of temperatures recorded in the layers of clay was found to go up and down in fine agreement with the records from the ice of Greenland and Antarctica. Researchers began to combine all these data in a single discussion.

The finest core of all was extracted by a French-Soviet team at the Soviet Vostok Station in Antarctica. It was a truly heroic feat of technology, wrestling with drills stuck a kilometer down at temperatures so low that a puff of breath fell to the ground in glittering crystals. Vostok was the most remote spot on the planet, supplied once a year by a train of vehicles that clawed across hundreds of kilometers of ice. Underfunded and threadbare, the station was fueled by the typically Russian combination of cigarettes, vodka, and stubborn persistence. In the late 1980s, when they reached bottom and hauled up their core, they found it stretched back over 400,000 years—through four complete cycles of glaciation.

For two decades, group after group had cut samples from cylinders of ice in hopes of measuring the level of CO_2 trapped within the tiny air bubbles. Every attempt had failed to give plausible results. Finally, in 1980, reliable methods were developed. The trick was to clean an ice sample scrupulously, crush it in a vacuum, and quickly measure what came out. The Vostok results were definite, unexpected, and momentous.

In each glacial period, the level of CO_2 in the atmosphere had been lower than during the warm periods in between—lower by as much as 50 percent. Nobody could explain what caused the level to soar and plunge so greatly as glacial periods came and went. The Vostok core tipped the balance in the greenhouse effect controversy, nailing down an emerging scientific consensus: the gas did indeed play a central role in climate change. This work fulfilled the old dream that studying the different climates of the past could be almost like putting the Earth on a laboratory table, switching conditions back and forth and observing the consequences.

The ice-core results answered the old persuasive objection to Milankovitch's orbital theory of ice ages: if their timing was set by variations in the sunlight falling on a given hemisphere, why didn't the Southern Hemisphere get warmer as the Northern Hemisphere cooled, and vice versa? The answer was that changes in atmospheric CO_2 (and methane, which likewise rose and fell) physically linked the two hemispheres, warming or cooling the planet as a whole. The findings also suggested one way in which the feeble shifts in sunlight of the 100,000-year cycle could raise and destroy continental ice sheets. It seemed there were indeed powerful feedbacks. Global warming and the natural emission of greenhouse gases somehow reinforced one another, severely amplifying any changes.

What was the feedback mechanism? The emission of gases from warmed-up wetlands and forests was only one of many candidates. Scientists suggested other possibilities, each more peculiar than the last. For example: in glacial ages the planet had bigger deserts and stronger winds; these put more dust into the air (the dust layers were visible in ice cores); the minerals in such dust could provide essential fertilization for ocean plankton, which perhaps would bloom prodigiously; the creatures would take up CO_2 from the air; when they died and drifted to the seafloor, the carbon would be buried with them, reducing the greenhouse effect; the glacial age would deepen. Or perhaps that was not what happened—there was no lack of ingenuity in finding other plausible mechanisms. Plainly there was a long way to go before anyone could explain the workings of major climate changes like the ice ages. After all, there was still no satisfactory analysis of the present climate.

For the present climate, as for the ice ages, dust lingered as a central puzzle. The arguments over whether aerosol particles warmed or cooled the planet were intricate and perplexing. For example, the historical research on volcanoes had turned up a distinct pattern of global cooling for a few years following each major eruption. But the particles of volcanic glass dropped out of the air within a few weeks. So how could they be responsible for such long-term effects? The atmosphere also rapidly rid itself of the mineral dust from soils degraded by human agriculture. Likewise it disposed of the carbon soot in smoke from factories and from slash-and-burn forest clearing. How could these make more than a temporary and local difference?

The answer was hidden in something else thrown into the air. Anyone looking at city smog—or smelling it—might guess that simple chemicals were a main component. The studies of smog that began in the 1950s brought a few scientists to look at chemicals in the air. They found that one of the most important molecules was sulfur dioxide, SO_2. Emitted profusely by volcanoes as well as by industries burning fossil fuels, SO_2 rises into the atmosphere and combines with water vapor to form minuscule droplets and crystals of sulfuric acid and other sulfates. In the early 1970s, when concerns about whether airplane emissions might harm the ozone layer had driven the U.S. government to conduct flights to study chemicals in the stratosphere, it was found that the most significant aerosols present were sulfuric acid and other sulfate particles. These lingered for years, reflecting and absorbing radiation. Did that make any difference? A hint came from the clouds covering the planet Venus. In the early 1970s, precise telescope observations identified the source of the haze that helped make the planet a greenhouse hell. The haze was mainly sulfates.

Haze on the Earth, outside the smoggy cities, was commonly assumed to be a "natural background" from soil particles, sea salt crystals, volcanoes, and so forth. That was challenged in 1976 by Bert Bolin and Robert Charlson. Analyzing air purity data collected by government agencies, they showed that sulfate aerosols measurably dimmed the sunlight across much of the eastern United States and Western Europe. Among all the aerosols arising from human activity, they calculated, sulfates played the biggest role for climate. The effect at that time seemed minor. But with consumption of fossil fuels rising, anyone who wanted to calculate the long-term future of climate would have to somehow take sulfates into account.

A few computer-modeling groups took up the challenge. Especially convincing was work published in 1978 by Hansen's group. They looked back to 1963, when the eruption of Mt. Agung in the Philippines had thrown some three million tons of sulfur into the stratosphere. A simplified model calculated that the sulfates should have made for cooling, and the numbers matched in all essential respects the global temperature changes actually observed in the mid-1960s. Contrary to what some scientists had argued, and unlike what happened on Venus, it seemed that the net effect of sulfates on Earth was to cool the surface.

Yet this was far from proving that the net effect of all human pollution would make for cooling. The arguments so far had only addressed how aerosols directly intercepted radiation. But since the 1960s, a few scientists had pointed out that this might not be their most important effect. New observations showed that under natural conditions, there were often too few nuclei to help water droplets coalesce into clouds. This was the kind of thing Walter Orr Roberts had talked about, when he had pointed to cirrus clouds evolving from jet contrails. That had been taken as a temporary, local effect. Now some wondered whether human emissions, by adding nuclei for water droplets, might increase cloudiness worldwide. If so, what would that mean for climate?

In 1977 Sean Twomey cast some light into these shadows. He showed that reflection of sunlight from clouds depends in a complex way on the number of nuclei. Depending on temperature, humidity, and the type and number of particles, aerosols might bring a thin mist, or a thick cloud, or rain followed by clear skies. So adding aerosols could either raise or lower cloud reflectivity, depending on a variety of factors. Moreover, while a cloud would reflect sunlight back into space, it would also intercept radiation coming up from below, causing a greenhouse effect. Extensive calculation was needed to figure whether the net effect of a particular sort of cloud was cooling or warming. Twomey calculated that overall, the effect of human aerosols should be to cool the Earth.

Other scientists paid little attention. The theory was too tricky for anyone to trust Twomey's calculations very far. And there were no convincing observations to pin down the equations—even the effects of seeding clouds with silver iodide smoke remained controversial despite decades of costly experiments. Hardly any measurements had been made of the mixture of particles and molecules that drifted in the atmosphere. Still less was known about the ways that the chemicals interacted with one another, although studies of smog showed that this was crucial. Few researchers cared to stake years of their professional lives on such hideously complex, perhaps insoluble, problems.

Yet the problems became steadily more unavoidable. Even in the Arctic, where the immense empty landscapes promised only pristine air, scientists were startled to find a visible haze of pollutants drifting up from industrial regions. They were coming to recognize that humans were now the dominant source of the atmosphere's sulfate aerosols. In 1987 a dramatic visible demonstration convinced many scientists that Twomey's calculations deserved respect. Satellite pictures of the oceans displayed persistent clouds above shipping lanes, a manifest response to smoke from the ships. Apparently aerosols did create enough clouds to significantly reflect sunlight. So what did that mean for the future climate? The few people dedicated to the troublesome theoretical and observational research were far from providing an answer.

Research moved ahead faster on an equally difficult, but more obviously central feature of the climate system. "We may find that the ocean plays a more important role than the atmosphere in climatic change," a panel of experts remarked in 1975.[12] The first generation of general circulation models had treated the oceans as if they were simply a wet surface, a featureless swamp. But ocean currents carried a huge amount of heat from the tropics to the poles. That was a key component of the engine of climate, a component that the GCMs—models of the general circulation of the atmosphere—had not tried to include.

Two obstacles kept modelers from handling the oceans in the same way as the atmosphere. First, while the atmosphere was measured daily in thousands of places, oceanographers had only occasional and scattered data. The sporadic expeditions, retrieving bottles of water here and there from kilometers down, were like a few blind men crawling about a vast prairie. Second, models of the atmosphere could bypass many difficulties by letting a simple equation or average number stand in for the complexities of a swirling storm; but analogous processes in the seas, the heavy decades-long sloshing of water about an ocean basin, had to be computed in full detail. The fastest computers of the 1970s lacked the capacity to calculate central features of the ocean system. They could not handle even something as fundamental and apparently simple as the vertical transport of heat from one layer to the next.

Oceanographers were coming to realize that crucial energy transfers were carried by a myriad of whorls of various sizes. At one extreme were microscopic circuits that shuttled heat down from the surface in some fashion nobody had yet figured out. At the other extreme were eddies bigger than Belgium that plowed through the seas for months. These colossal slow whorls were only discovered in the 1970s, thanks to an international study of the North Atlantic carried out by six ships and two aircraft. To the amazement of oceanographers, most of the energy in the ocean system was carried about by these eddies, not by ocean-spanning currents like the Gulf Stream. Calculating all the big and little whorls, like calculating individual clouds, was far beyond the reach of the fastest computer. Modelers had to work out parameters to summarize the main effects, only this time for objects that were much harder to observe and less understood than clouds. Even with gross simplifications, to get anything halfway realistic took many more numerical computations than for the atmosphere.

Suki Manabe had shouldered the task in collaboration with Kirk Bryan, an oceanographer with training in meteorology who was brought into the Princeton group back in 1961 to build a stand-alone numerical model of an ocean. The two had teamed up to construct a model that coupled this ocean model to Manabe's atmospheric GCM. Manabe's winds and rain would help drive Bryan's ocean currents, while in return Bryan's sea-surface temperatures and evaporation would help determine the circulation of Manabe's atmosphere. In 1968 they completed a heroic computer run, some 1100 hours long (totaling over 12 full days devoted to the atmosphere and 33 to the ocean).

Bryan wrote modestly that "in one sense the . . . experiment is a failure."[13] Even after running for a simulated century, the deep-ocean circulation had not nearly reached equilibrium. It was not clear what the final climate solution would look like. Yet it was a great success just to get a linked ocean-atmosphere computation that was at least starting to settle into an equilibrium. The result looked like a real planet—not our Earth, for in place of geography they used a radically simplified geometrical sketch, but with plausible ocean currents, trade winds, deserts, snow cover, and so forth.

Unlike our actual Earth, so poorly observed, in the simulation one could see exactly how air, water, and energy were moved about.

Following up, in 1975 Manabe and Bryan published results from the first coupled ocean-atmosphere GCM that had a roughly Earth-like geography. The supercomputer ran for fifty straight days, simulating movements of air and sea over nearly three centuries. At the end their simulated world-ocean still failed to show a full circulation. But the results were getting close enough to reality to encourage them to push ahead. Meanwhile Washington's team in Boulder developed another ocean model, based on Bryan's, and coupled it to their own quite different GCM. Their results resembled Manabe and Bryan's, a gratifying confirmation.

Ocean modeling was becoming a recognized specialty. A research program that had once seemed "a lonely frontier like a camp of the Lewis and Clark Expedition," as an ocean modeler recalled in 1975, took on "more of the character of a Colorado gold camp."[14] One reason was the breathtaking advances in computers. Equally important was a splendid addition to the limited stock of oceanographic data. In a major project that the U.S. government funded in the 1970s, the Geochemical Ocean Sections Study (GEOSECS), teams of researchers sampled seawater at many points. Their chief interest was the radioactive carbon, tritium, and other debris spewed into the atmosphere by nuclear bomb tests in the late 1950s. The fallout had landed on the ocean surface around the world and was gradually being carried into the depths. Thanks to its radioactivity, even the most minute traces could be detected. The bomb fallout "tracers" gave enough information to map accurately, for the first time, all the main features of the three-dimensional ocean circulation. At last modelers had a realistic target to aim at.

Various teams improved their ocean-atmosphere models through the 1980s, occasionally checking how they reacted to increased levels of CO_2. The results, for all their limitations, said something about the predictions of the earlier atmosphere-only

GCMs. It turned out, as expected, that the oceans would delay the appearance of global warming for a few decades by soaking up heat. As Hansen's group warned, a policy of "wait and see" might be wrongheaded, for a temperature rise in the atmosphere might not become apparent until much more greenhouse effect warming was inevitable.[15] Aside from that, linking a somewhat realistic ocean to GCMs did not turn up anything to alter the predictions in hand for future warming.

There was a problem, however, which few appreciated. The computer models showed a steady, gradual change of climate as CO_2 increased. But in their very structure, the models were designed to show only smooth changes. In the real world, when you push on something steadily it may remain in place for a while, then move with a jerk. Since the 1960s scientists had suspected that this sometimes happened to the climate system, and in the 1980s disturbing new evidence confirmed it.

There had been hints of rapid climate change in the long core drilled in the 1960s at Camp Century in Greenland, but a single record could be subject to all kinds of accidental errors. Dansgaard's group built a new drill and went to a second location, some 1,400 kilometers distant from Camp Century, where they extracted gleaming cylinders of ice ten centimeters in diameter and in total over two kilometers long. They cut out 67,000 samples, and in each sample analyzed the ratios of oxygen isotopes. The temperature record showed jumps that corresponded closely to the jumps at Camp Century.

In 1984 Dansgaard reported that the most prominent of the "violent" changes corresponded to the Younger Dryas oscillation, "a dramatic cooling of rather short duration, perhaps only a few hundred years."[16] Corroboration came from a group working under Hans Oeschger. An ice-drilling pioneer, Oeschger was now analyzing layers of lake-bed clay near his home in Bern, Switzerland. That was far indeed from Greenland, but his group found "drastic climatic changes" that neatly matched the ice record.[17]

Many felt that such large changes had to be regional, perhaps affecting the North Atlantic and Europe but not the whole planet. A look at North American and Antarctic records did not find the same features. Yet as ice drillers improved their techniques, they found, to everyone's surprise, large steps not only in temperature but also in the CO_2 concentration. Since the gas circulates through the atmosphere in a matter of months, the steps seemed to reflect abrupt worldwide changes. Oeschger was particularly struck by a rapid rise of CO_2 that others reported finding in Greenland ice cores at the end of the last ice age.

The main reservoir of carbon was the oceans, so that was the first place to think about. In 1982 Broecker visited Oeschger's group in Bern and explained current ideas about the North Atlantic circulation. Oeschger pondered how the carbon balance of the oceans might be changed, but could not come up with a convincing mechanism. In fact, scientists later realized that the rapid variations seen in the ice cores were merely an artifact. They did not reflect changes in atmospheric CO_2, but only changes in the ice's acidity due to dust layers (something had indeed changed swiftly, but not necessarily the CO_2 level). No matter: the error had served a good purpose, for Oeschger's speculations had set Broecker to thinking.

Ever since the days when he had trudged around fossil lake basins in Nevada for his doctoral thesis, Broecker had been interested in sudden climate shifts. The reported sudden jumps of CO_2 in Greenland ice cores stimulated him to put this interest into conjunction with his oceanographic interests. The result was a surprising and important calculation. The key was what Broecker later described as a "great conveyor belt" of seawater carrying heat northward. The gross properties of the circulation had been laid out a decade earlier by the GEOSECS survey of radioactive tracers.

But it was only now, as Broecker and others worked through the numbers in enough detail to make crude computational models, that they fully grasped what was happening. The vast mass of water that gradually creeps northward near the surface of the Atlantic is as important in carrying heat as the familiar and visible Gulf Stream. The energy carried to the neighborhood of Iceland was "staggering," Broecker realized—nearly a third as much as the Sun sheds upon the entire North Atlantic.[18] If something were to shut down the conveyor, climate would change across much of the Northern Hemisphere. England is as far north as Labrador, and but for the heat in the ocean, it would be as cold. In 1985 Broecker and two colleagues published a paper titled, "Does the Ocean-Atmosphere System Have More than One Stable Mode of Operation?" Their answer was yes: the great conveyor belt could easily shut down.

In one sense this was no discovery. It was an extension of the conjecture that Chamberlin had offered at the start of the century, that the circulation could shut down if the North Atlantic surface water became less salty (Chapter 1). Few scientific "discoveries" are wholly new. An idea moves from a casual speculation to a discovery when something makes it look truly plausible. Broecker did that by calculating solid numbers. He also pointed out evidence that such a shutdown had actually happened. Geological studies showed that at the end of the last glacial epoch, as the North American ice sheet melted, it had dammed up a huge lake; when this was suddenly released, it had sent a colossal surge of fresh water into the ocean. It seemed likely that this had caused a shutdown and cooling. The timing was right: just at the start of the Younger Dryas.

Computer modeling teams now took a closer look at the thermohaline circulation of the oceans, the world-spanning movement of seawater in which differences in heat and salinity drove a vast overturning in the North Atlantic. They found that it was indeed precarious. The melting of a continental ice sheet was not necessary to stop the circulation. The balance might be shifted, for example, by additional fresh-water rainfall—which might accompany global warming. In 1985 Bryan and a collaborator tried out a coupled atmosphere-ocean model with a CO_2 level four times higher than at present. They found signs that the ocean circulation could come to a halt. Three years later, Manabe and another collaborator found that the ocean-atmosphere system was so delicately balanced that even at present CO_2 levels, if it got into an alternative state in which the steady flow of warm water into the North Atlantic was halted, it would stay that way.

There was reason to believe a shutdown could happen swiftly. In many regions the consequences for climate would be spectacular. Broecker was foremost in taking this disagreeable news to the public. In 1987 he wrote that we had been treating the greenhouse effect as a "cocktail hour curiosity," but now "we must view it as a threat to human beings and wildlife." The climate system was a capricious beast, he said, and we were poking it with a sharp stick.[19]

CHAPTER SEVEN

BREAKING INTO POLITICS

Around 1966 Roger Revelle gave a lecture about the Earth's future to students at Harvard University. Among the undergraduates was a senator's son: Albert Gore, Jr. When Revelle displayed Keeling's curve of CO_2 in the atmosphere, which at this point showed an eight-year climb, Gore was deeply impressed. The prospect of greenhouse warming came as a shock to him, he later recalled, exploding his childhood assumption that "the Earth is so vast and nature so powerful that nothing we do can have any major or lasting effect on the normal functioning of its natural systems."[1]

By 1981 Al Gore was a Representative in Congress, and better prepared than anyone to haul climate change from the halls of scientific discussion onto the stage of political controversy. Over the years he had kept abreast of the technical issues as they developed, and he shared the concern about global warming as it grew among scientists. No doubt Gore also saw a political opening. As a champion of environmental issues, he could display leadership in one of the few areas where the newly installed Republican administration's policies disturbed a large majority of voters.

Ronald Reagan had assumed the presidency with an administration that openly scorned environmental worries, global warming included. Many conservatives lumped all environmental concerns together as the rantings of liberals hostile to business, a Trojan horse for the expansion of government regulation and secular values. The recently established National Climate Program Office found itself serving, as an observer put it, as "an outpost in enemy territory."[2] The new administration laid plans to slash funding for CO_2 studies in particular, deeming such research unnecessary. They even targeted support for monitoring the level of CO_2 in the atmosphere—the indefatigable Keeling's rigorous measurements, which now showed more than two decades of relentless rise.

Gore and a few other representatives decided to embarrass the administration with Congressional hearings on the proposed cuts. In the hearings, it is doubtful that any representatives were swayed by the technical explanations of the greenhouse effect. But some newspapers noticed the testimony by persuasive scientists like Revelle and Schneider. As an aide close to the political process put it, "the popular media is the most potent way of convincing a member of Congress that he should pay attention to scientific issues." Politicians did not read scientific journals, but relied on the press as the "prime detector of the public's fears."[3]

When it came to deciding what scientific developments were "news," American journalists tended to take their cues from the *New York Times*. The editors of the *Times* followed the advice of their veteran science writer, Walter Sullivan. A lanky and amiable reporter, Sullivan had frequented meetings of geophysicists ever since the International Geophysical Year of 1957, cultivating a number of trusted advisers. On the subject of climate he began listening to scientists like Schneider, Hansen, and a few others who were alarmed about global warming and determined to attract attention to the issue.

In 1981, for example, Hansen sent Sullivan a scientific report he was about to publish, the one announcing that the planet was getting noticeably warmer (Chapter 6). For the first time the greenhouse effect made page one of the *New York Times*. Sullivan threatened the world with an unprecedented warming, which might cause a disastrous rise of sea level. The newspaper followed up with an editorial, declaring that while the greenhouse effect was "still too uncertain to warrant total alteration of energy policy," it was "no longer unimaginable" that a radical policy change might become necessary.[4] The Department of Energy responded by reneging on funding they had promised Hansen, and he had to lay off five people from his institute.

Everything connected with atmospheric change had become politically sensitive. Thus scientists were reporting that "acid rain" was ravaging forests (and even the paint on houses) thousands of miles downwind from smokestacks that emitted sulfates. When environmentalists demanded restrictions, the coal industry counterattacked with its own scientists and advertising to promote an image of benign economic progress which could never cause long-range damage.

Another important example was a controversy that erupted on Halloween 1983. That day a group of respected atmospheric scientists issued a carefully orchestrated announcement of a novel risk of climate catastrophe, as a likely consequence of nuclear war. Like so much in geophysics, the idea had its origins in an unexpected place. In 1980, geologist Walter Alvarez and his physicist father, Luis Alvarez, had suggested that the extinction of the dinosaurs had been caused by an asteroid that struck the Earth. The clouds of dust and smoke would have obscured the atmosphere for years, freezing plants and animals. This image of death from the skies, apocalyptic explosion, and worldwide extinction resonated with the fear of nuclear war that lay deep within every mind in this time of revived Cold War rhetoric. A few scientists developed computer models to test the effects of an atmosphere filled with smoke and dust, and it occurred to them to apply these models to nuclear war.

They calculated that after an exchange of hydrogen bombs, the sooty smoke from burning cities could bring on a global "nuclear winter"—months or even years of cold and dark that could threaten the survival of humankind. This was the subject of the 1983 press conference. In the fore was Carl Sagan, whose fame— much more as an astronomy popularizer than as an atmospheric scientist—could attract television cameras. Didn't the calculations prove that launching a nuclear attack, even if the other side never fired back, would be literally suicidal? So maintained Sagan and his allies, with a frankly political aim. They hoped to reinforce a public movement that was demanding that the United States reduce its inventory of bombs. Meanwhile they added another layer to public imagination of global climate catastrophe.

Other scientists questioned the scientific reasoning, and the Reagan administration heaped scorn on its critics. A sharp partisan debate followed. With high stakes in play, reasoned public discussion was increasingly drowned out by the shouts of polarized combat. Computer calculations of the effects of dust and the fragility of the atmosphere had become inescapably entangled in national politics. If you knew a person's views on nuclear disarmament, you could probably predict what the person thought about the nuclear winter prediction. And if you knew a person's views on government regulation, you could probably predict what the person thought about predictions of global warming.

Most scientists struggled to stand apart from policy debates, even when they were called upon to state what the scientific community reliably knew and could agree upon. The traditional authority for providing such advice was the U.S. National Academy of Sciences. In 1980 Congress had asked the Academy to carry out a comprehensive study on the impacts of rising CO_2. The Academy appointed a panel of leading experts, and in 1983, following a sustained effort to work out a consensus, the panel reported. The scientists said they were "deeply concerned" about the environmental changes that they expected a temperature rise would bring. And they pointed out that "we may get into trouble in ways that we have barely imagined"—for example, if warming released methane from seabed sediments. But on the whole, the panel was cautiously reassuring. They said the warming would probably not be very severe. And a degree or two of temperature change was something people in the past had managed to get through well enough. They advised against any immediate policy changes, such as trying to restrict use of fossil fuels. The Academy's chief recommendation was that before doing anything, the government should fund vigilant monitoring and other studies—that is, More Money Should Be Spent on Research.[5]

Three days later, the Environmental Protection Agency released a report of its own on the greenhouse effect. The science was mostly the same, but the tone of the EPA's conclusions was more anxious. A ban on fossil fuels seemed out of the question on both economic and political grounds. So the panel saw no practicable way to prevent a temperature rise. It could be a big rise, within a few decades, with potentially "catastrophic" consequences.[6] The *New York Times* took notice in a front-page story (Oct. 18, 1983). This EPA report was the first time a Federal agency declared that global warming was, as the reporter put it, "not a theoretical problem but a threat whose effects will be felt within a few years."

Administration officials criticized the EPA report as alarmist, pointing to the more reassuring Academy report. Here was a tale of battling views, just what journalists needed to make a lively story. It spread through the newspapers and even got onto national television. The controversy, piled on top of Congressional hearings and the efforts of scientist publicists, alerted a sizable fraction of citizens and politicians to the prediction that was shared by both reports. It was official: global warming might well be coming. Climate scientists found themselves in demand to give tutorials to journalists, government agency officials, and even groups of senators, who would sit obediently for hours of lecturing on greenhouse gases and computer models.

If global warming was coming, what would be the consequences? Some things seemed quite certain on basic physical principles, confirmed by computer models. Science reporters took care that the public, if they read about the topic at all, understood that an *average* global warming of three degrees did not mean that the thermometer would be exactly three degrees higher, everywhere, every day. In a warmer world, some regions might not be affected much. Other regions would suffer unprecedented heat waves, sometimes deadly ones. Less obviously, the air would hold more moisture, so the water cycle might intensify. Some regions would have more storms, or greater floods, or worse droughts.

And the sea level could rise. Researchers had not been able to dismiss worries about a collapse of the West Antarctic Ice Sheet. Too little was known about the way ice behaved for experts to agree on any firm conclusion. Some studies saw the possibility of a collapse and sea-level rise of two or three meters (6 to 10 feet) by the year 2100. Most experts disagreed, calculating that this could not possibly happen so soon. Yet a gradual discharge of ice over the following centuries was possible, and that could place a heavy burden on human society.

A smaller, but significant, rise of sea level within the coming century was expected for another reason. Water expands when heated. The consequences may seem obvious, but amid all the talk of melting glaciers, for decades nobody seems to have given a thought to other simple effects. Eventually, in 1982, two groups separately calculated that the global warming observed since the mid-nineteenth century must have raised the sea level significantly by plain thermal expansion of the upper ocean layers. There had indeed been a rise of 10 or 20 centimeters (ten times faster than the average change in earlier millennia). Expansion could not account for all of it, and the scientists figured the rest came from melting glaciers—most of the world's small mountain glaciers were in fact shrinking. In the twenty-first century, they warned, rising tides would erode shore-lines back hundreds of feet. Salt water would invade estuaries. Entire populations would flee from storm surges.

So it was widely agreed that global warming could be a threat, and that the proper response was to study it. Weary of the issue and distracted by more urgent matters, the media and public turned their attention elsewhere. Funding for the environmental sciences was not drastically cut after all, but neither was it expanded. During the 1980s, the U.S. government spent barely $50 million per year for research directly focused on global warming. It was a trivial sum compared with many other research programs. Other nations did not take up the slack. Through Western Europe to the Soviet Union, governments scarcely increased their spending on science in general and climate science in particular.

Out of the money that government agencies did have available, they diverted part into a new type of study: the social and economic impacts of climate change. What would it mean for agriculture, forestry, the spread of tropical diseases, and so forth? Answers were not easy to come by. A few scientists claimed that more CO_2 would be a good thing, on the whole. Especially in frigid Russia, many looked forward to warming, or anyway the first few degrees of warming. But as studies were refined over following decades, they found mainly negative impacts over the long run. It gradually became apparent that even a degree or two of warming of the oceans could devastate many of the world's coral reefs, that tropical diseases would invade new territory, and so forth.

Such research required a broadly multidisciplinary approach. Climate experts began to interact with experts and policy-makers in agriculture, economics, and many other fields. Researchers no longer spoke of studying "climates" in the old sense of regional weather patterns, but rather "the climate system" of the whole planet, involving everything from minerals to microbes, not to mention the rapidly evolving human activities. More and more often, specialists in diverse fields met in committees and panels that advised national and international research programs.

Universities and other institutions likewise encouraged coalitions of research groups in a variety of fields. Specialists in the quirks of the stratosphere, volcanoes, the oceans, and even biology, not to mention the mathematics of computation, found themselves sharing the same funding agencies, institutions, and even buildings. Also growing common were scientific meetings devoted to one or another interdisciplinary topic. Collaboration was a powerful trend not only in geophysics but in all the sciences. As research problems spanned ever more complexities, scientists with different types of expertise exchanged ideas and data, or worked directly together for months if not years. Before 1940, very few of the papers published on climate had two authors. By the 1980s, the majority of papers had more than one author, and papers with seven or eight authors were no longer surprising.

None of this entirely solved the problem of fragmentation. The more the research enterprise grew, the more narrowly scientists were driven to specialize. Meanwhile the imperatives of administration maintained boundaries between academic disciplines and between the government agencies and organizations that supported them. Most scientific papers continued to be published in journals dedicated to a particular field, like the meteorologists' *Journal of the Atmospheric Sciences* or the paleontologists' *Quaternary Research*. At the same time, every scientist read *Science* and *Nature*. These comprehensive journals competed with one another for the most important papers in every scientific field, including those connected with climate change. Both journals also published expert reviews and staff commentaries that helped keep scientists up to date on developments outside their own field.

The dispersion of research funding among several agencies of the U.S. government was partly solved by an Earth System Sciences Committee that NASA set up in 1983. The committee members struck bargains among agencies and disciplines, forging a common front. They managed to organize a large-scale, interdisciplinary initiative to study global change, with full cooperation from the agencies. After fighting out their differences among themselves, the leaders agreed on a short list of top-priority programs and put the weight of their joint prestige behind it. The administration's budget-makers and Congress, pleased to see a well-coordinated effort, opened their pockets. Meanwhile the committee worked out cooperation with scientists around the world. For climate research was finally getting its own international organization, strong enough to win support from the various individual nations.

The hundreds of scientists and government officials who assembled at the World Climate Conference held in Geneva in 1979 (Chapter 5) called for an international structure set up specifically for climate research. The government representatives in the WMO and the scientific leaders in ICSU took the advice and joined to launch a World Climate Research Program (WCRP). It took over the portion of the old Global Atmospheric Research Program that had been concerned with climate change, including a small staff in Geneva and an independent scientific planning committee. The WCRP had various branches, each known mainly by an acronym. (This mode of naming was emblematic of organizations with distinct if temporary identities made up of independent components.) For example, an International Satellite Cloud Climatology Project (ISCCP) collected streams of raw data from the weather satellites of several nations and fed the data through a variety of government and university groups for processing, analysis, and archiving. These U.N.-sponsored efforts were only one strand, although the central one, in a tangle of national, bilateral, and multi-national climate initiatives. Countless organizations were now seeking to be part of the action.

Of course none of this proliferation was really the work of abstract entities. It was brought about by a few human beings, scientists and officials dedicated to international and environmental interests. They blurred the distinction between governmental and private initiatives as they organized quasi-official international meetings. The indispensable man was Bert Bolin, who kept busy chairing meetings, editing reports, and promoting the establishment of panels. Along with his outstanding personal abilities as a scientist, executive, and diplomat, Bolin benefited from his position at the University of Stockholm in Sweden, traditionally neutral territory.

The most important initiative was a series of invitational meetings for meteorologists held at Villach, Austria, through the 1980s. The 1985 Villach conference was a turning point. The assembled experts announced an international consensus: "in the first half of the next century a rise of global mean temperature could occur which is greater than any in man's history." As usual, the scientists called for more research, but they also took a relatively activist stance. Governments must take action. "While some warming of climate now appears inevitable due to past actions, the rate and degree of future warming could be profoundly affected by governmental policies."[7]

Villach and other international meetings, along with similar consensus-building studies on climate change carried out by national bodies such as the U.S. National Academy of Sciences, crystallized a set of beliefs and attitudes among climate scientists. As one science writer reported after a series of interviews, "By the second half of the 1980s, many experts were frantic to persuade the world of what was about to happen."[8]

Human motivation is seldom simple, and behind the emotional commitment of scientists lay more than dry evaluation of data. Adding to their concern about global warming was the normal tendency of people to perceive their own fields as vitally important, with the corollary that funds should be generously awarded for their work and for their students and colleagues. The scientists found allies among administrators in national and international bureaucracies, persuading many that the world faced a serious problem. That reinforced the normal inclination of officials to extol the importance of their areas of responsibility, and to seek greater budgets and broader powers. Whenever there is evidence that something needs to be done, those who stand to profit from the doing will be especially quick to accept the evidence and argue for policy changes. The few politicians who joined, like Gore, likewise found that personal convictions fitted with opportunities for career advancement.

The only reliable way to sort through the human motives and determine what policy action was really needed was to seek rigorous scientific conclusions. While a few scientists and officials tentatively proposed policy changes, many more were pushing for still larger international research projects. The WCRP was all to the good, but it stuck too narrowly to meteorology. Around 1983 various organizations collaborated under ICSU to draw together all the geophysical and biological sciences in an International Geosphere-Biosphere Program (IGBP). Starting up in 1986, the IGBP built a large structure of committees, panels, and working groups that fostered interdisciplinary connections. The drawback, as Schneider pointed out, was a feeling that "an IGBP should be in the business of measuring or modeling everything at once from the mantle of the Earth to the center of the Sun!"[9]

Research did result in a big policy breakthrough in the late 1980s, although not for climate. Ever since the spray-can controversy of the mid-1970s, scientists had worried about the destruction of ozone in the stratosphere. In 1985 this led 20 nations to sign the Vienna Convention for the Protection of the Ozone Layer. The document was only a toothless expression of hopes, but it established a framework. The framework became useful almost at once. In 1985 a British group announced their discovery of a "hole" in the ozone layer over Antarctica. The apparent culprit was again CFCs, banned from American spray cans but still widely produced around the world for a variety of purposes. Inevitably a new controversy began, for industrial groups automatically denied any of their products could be risky, and Reagan administration officials reflexively backed the industries against hostile environmentalists.

The denials were short-lived. Within two years, new theories of how the chemicals could destroy ozone, confirmed by daring flights over Antarctica, convinced the experts. The immediate threat of increased skin cancers and other damage to people and biological systems shocked officials. Meanwhile magazine and television images of the ominous map of ozone loss carried the message to the public. Most people lumped together all forms of potential atmospheric harms, with the threat to the ozone coming on top of greenhouse gases, smog, acid rain, and so forth. Politicians were forced to respond. In the epochal 1987 Montreal Protocol of the Vienna Convention, the world's governments formally pledged to restrict emission of specific ozone-damaging chemicals.

This was not the first international agreement to restrict pollution in response to scientific advice. In 1979, for example, the nations of Western Europe had adopted a convention to address acid rain, pledging to study their sulfate emissions and impose limits on them. The Montreal Protocol set an even stricter standard for international cooperation and national self-restraint. Over the following decade, it had great success in reducing emissions of CFCs. Although essential for protecting the ozone layer, the suppression of these chemicals was not much help for climate. Some of the chemicals that industry substituted for CFCs were themselves greenhouse gases. So was ozone, and keeping it in the stratosphere would add a bit to global warming.

At the same time, many environmentalists hoped that the precedent set by the Montreal Protocol could show the way to restrictions on greenhouse gases. Industrialists and ideologues had opposed such regulation as an insufferable economic drag. But for the regulation of CFCs, like the regulation of sulfate emissions and a variety of other pollutants, it turned out that market-oriented mechanisms could be devised to do the job cheaply—over the long run at a net *savings* to the global economy.

The success at Montreal was followed up the next year, 1988, with a "World Conference on the Changing Atmosphere: Implications for Global Security," nicknamed the Toronto conference. The planning came out of the workshops initiated by the 1985 Villach conference. Toronto was a meeting by invitation of scientist experts—not official government representatives, who would have had a much harder time reaching a consensus. The Toronto conference report concluded that human pollution of the atmosphere was already causing harm and should be faced without delay. For the first time, a group of prestigious scientists called on the world's governments to set strict, specific targets for reducing greenhouse gas emissions. That was the Montreal Protocol model: set targets internationally and let governments come up with their own policies to meet them. By 2005, said the experts, emissions ought to be pushed some 20 percent below the 1988 level.

Up to this point, the summer of 1988, global warming had been generally below the threshold of public attention. The reports that the 1980s were the hottest years on record had barely made it into the inside pages of newspapers. A majority of people were not even aware of the problem. Those who had heard about global warming mostly saw it as a gradual matter, something that the next generation might or might not need to worry about. Yet a shift of views had been prepared by the ozone hole, acid rain, and other atmospheric pollution stories, by a decade of agitation on these and many other environmental issues, and by the slow turn of scientific opinion toward strong concern about global warming. To ignite the worries, only a match was needed. This is often the case for matters of intellectual concern. No matter how much pressure builds up among concerned experts, some trigger is needed to produce an explosion of public attention.

The break came in the summer of 1988. A series of heat waves and droughts, the worst since the Dust Bowl of the 1930s, devastated many regions of the United States. Cover articles in news magazines, lead stories on television news programs, and countless newspaper columns offered dramatic images of parched farmlands, sweltering cities, a "super hurricane," and the worst forest fires of the century. Reporters asked: were all these caused by the greenhouse effect? Scientists knew that no individual weather event could be traced to global warming. But simply from endless repetition of the question, many people became half convinced that our pollution was indeed to blame for it all.

In the middle of this Hansen raised the stakes with deliberate intent. By arrangement with Senator Timothy Wirth he testified at a Congressional hearing in late June, deliberately choosing the summer, although that was hardly a normal time for politicians who sought attention. Outside the room the temperature that day reached a record high. Inside, Hansen said he could state "with 99 percent confidence" that there was a long-term warming trend under way, and he strongly suspected that the greenhouse effect was to blame. He and like-minded scientists testified that global warming could bring more frequent storms and floods as well as life-threatening heat waves. Talking with reporters afterward, Hansen said it was time to "stop waffling, and say that the evidence is pretty strong that the greenhouse effect is here."[10] As the heat waves and drought continued, reporters descended unexpectedly upon the Toronto conference and prominently reported its alarming conclusions. The story was no longer a scientific abstraction about an atmospheric phenomenon: it was about a present danger to everyone, from elderly people struck down by heat to the owners of beach houses. Images of blasted crops and burning forests seemed like a warning signal, a visible preview for what the future might hold.

The media coverage was so extensive that, according to a 1989 poll, 79 percent of Americans recalled having heard or read about the greenhouse effect. This was a big jump from the 38 percent who had heard about it in 1981, and an extraordinarily high level of public awareness for any scientific phenomenon. Most of these citizens recognized that "greenhouse effect" meant the threat of global warming, and most thought they would live to experience climate changes.

The environmental movement, which had taken only an occasional interest in global warming, now took it up as a main cause. Groups that had other reasons for preserving tropical forests, promoting energy conservation, slowing population growth, or reducing air pollution could make common cause as they offered their various ways to reduce emissions of CO_2. Adding their voices to the chorus were people who looked for arguments to weaken the prestige of large corporations, and people ready to scold the public for its wastefulness. For better or worse, global warming became firmly identified as a "green" issue.

In the long perspective, it was an extraordinary novelty that such a thing became a political question at all. Global warming was invisible, no more than a possibility, and not even a current possibility but something predicted to emerge only after decades or more. The prediction was based on complex reasoning and data that only a scientist could understand. It was a remarkable advance for humanity that such a thing could be a subject of widespread and vehement debate. Discourse had in many ways grown more sophisticated. That was perhaps because of the steady accumulation of knowledge, and also, perhaps, because the general public in wealthy countries had become somewhat educated (a higher proportion of young people were now going to college than had gone to high school at the turn of the century). Stable times, and the unexpected addition of decades to the average lifespan, encouraged people to plan farther into the future than in earlier times.

Politicians at the highest level began to pay attention to greenhouse gases. One main reason was the exceptional heat and drought that caused so much public concern in the United States, the nation whose cooperation was indispensable for any agreement (in the capital itself, the summer of 1988 was the hottest on record). But officials were also impressed by the insistent warnings of leading scientists. In the United Kingdom, Prime Minister Margaret Thatcher gave global warming official standing when she described it as a key issue. Attention from the politically powerful "Greens" in Germany and elsewhere in continental Europe added to the issue's legitimacy. What had begun as a research puzzle had become a serious international diplomatic issue.

Negotiations and regulations would be shaped (at least in part) by scientific findings. That was a heavy burden for climate scientists. Starting around 1988, climate research topics became far more prominent in the scientific community itself. The volume of publications soared, and there were so many workshops and conferences that nobody could attend more than a fraction of them. This sudden growth of scientific attention was probably partly due to the step-up in public concern: anyone studying the topic would now get a better hearing when requesting funds, recruiting students, and publishing. In turn, the new emphasis on climate research was reflected by science journalists, who took their cues from the scientific community and passed its views on to the public.

Yet there were still only a few hundred people in the world who devoted themselves full-time to the study of climate change. And they were still dispersed among many fields. Somehow their expertise had to be assembled to answer the urgent policy questions. The reports issued by brief conferences could not command thorough respect, and they did not commit any particular group to follow up.

Programs like the IGBP were designed only to promote an array of research projects. Some new kind of institution was needed.

Conservatives and skeptics in the United States administration might have been expected to oppose the creation of a prestigious body to study climate change. But they distrusted still more the system of international panels of independent scientists that had been driving the issue. If the process continued in the same fashion, the skeptics warned, future groups might make radical environmentalist pronouncements. Better to form a new system under the control of government representatives. Besides, a complex and lengthy study process would delay any move to take concrete steps to restrain emissions.

Responding to all these pressures, in 1988 the WMO and other United Nations environmental agencies created an Intergovernmental Panel on Climate Change (IPCC). Unlike earlier bodies, the IPCC was composed mainly of representatives of the world's governments—people who had strong links to national laboratories, meteorological offices, science agencies, and so forth. It was neither a strictly scientific nor a strictly political body, but a unique hybrid.

Most people were scarcely aware of the central factor in these international initiatives: the worldwide advance of democracy. It is too easy to overlook the obvious fact that international organizations are governed in a democratic fashion, with vigorous free debate and votes in councils. Often, as in the IPCC, decisions are made by a negotiated consensus in a spirit of equality, mutual accommodation, and commitment to the community process (a seldom appreciated part of the democratic culture). If we tried to make a diagram of the organizations that dealt with climate change, we would not draw an authoritarian tree of hierarchical command, but a spaghetti tangle of cross-linked, quasi-independent committees.

It is an important but little known fact that such organizations were created primarily by governments that felt comfortable with such mechanisms at home, that is, democratic governments. Happily, the number of nations under democratic governance increased dramatically during the twentieth century, and by the end of the century they were predominant. Therefore democratically based international institutions proliferated, exerting an ever stronger influence in world affairs.[11] This was seen in all areas of human endeavor, but it often came first in science, internationally minded since its origins. The democratization of international politics was the scarcely noticed foundation upon which the IPCC and its fellow organizations took their stand.

It worked both ways. The international organization of climate studies helped fulfill some of the hopes of those who, in the aftermath of the Second World War, had worked to build an open and cooperative world order. If the IPCC was the outstanding example, in other areas, ranging from disease control to fisheries, panels of scientists were becoming a new voice in world affairs. Independent of nationalities, they wielded increasing power by claiming dominion over views about the actual state of the world—shaping perceptions of reality itself. Such a transnational scientific influence on policy matched dreams held by liberals since the nineteenth century. It awoke corresponding suspicions in the enemies of liberalism.

CHAPTER EIGHT

THE DISCOVERY CONFIRMED

The closer an account of events moves toward the present, the less can it be called "history," and the more it looks like something else—perhaps journalism. The special virtues we seek in a work of history, the long-view perspectives and objective analysis, dwindle. Writers and readers find it hard to pick out which recent developments will really matter in the long run. Worse, opinions about present-day controversies infect views of the recent past with special virulence. Therefore be wary: this concluding chapter can be only a preliminary sketch.

By the late 1980s, well-informed people understood that the climate change issue could not be handled in either of the two easiest ways. Scientists were not going to prove that there was nothing to worry about. Nor were they about to prove exactly how climate would change, and tell what should be done about it. Spending more money on research would no longer be a sufficient response (not that enough had ever been spent). For the scientists were not limited by the sort of simple ignorance that could be overcome with clever studies. A medical researcher can find the effects of a drug by giving a thousand patients one pill and another thousand patients a different one, but climate scientists did not have two Earths with different levels of greenhouse gases to compare. The best they could do was build elaborate computer models and vary the numbers that represented the level of gases. That hardly seemed a convincing way to tell the civilized world how it should reorganize the way everyone lived.

Of course, people make all their important decisions in uncertainty; every social policy and business plan is based on guesswork. But global warming would never have been an issue at all except for the scientists. Somehow scientists would now have to give the world practical advice, yet without abandoning the commitment to strict rules of evidence and reasoning that made them scientists in the first place.

The Intergovernmental Panel on Climate Change set to work under the judicious chairmanship of Bert Bolin. Unlike all the earlier national academy panels, conferences, and other bodies, the IPCC brought together people who spoke not just as science experts, but as official representatives of their respective governments. With all the world's important climate scientists and governments participating, the IPCC quickly established itself as the principal source of scientific advice to policy-makers.

The IPCC's method was to set up independent working groups to address each of the various scientific issues. Experts drafted reports drawing on the latest studies, and these were debated at length in workshops. Through 1989 the IPCC scientists, 170 of them in a dozen workshops, worked hard and long to craft statements that nobody could fault on scientific grounds. The draft reports next went through a process of review, gathering comments from virtually every climate expert in the world. The scientists found it easier than they had expected to achieve a consensus. But any final conclusions had to be endorsed by a consensus of government delegates, many of whom were not scientists.

Among the governments, the most eloquent and passionate in arguing for strong statements were representatives of small island nations, who had learned that rising sea levels could erase their territories from the map. Far more powerful were the oil, coal, and automobile industries, represented by governments of nations living off fossil fuels, like Saudi Arabia. The negotiations were intense, and only the fear of an embarrassing collapse pushed people through the grueling sessions to grudging agreement. Under pressure from the industrial interests, as well as from the mandate to make only statements that virtually every knowledgeable scientist could endorse, the IPCC's consensus statements were highly qualified and cautious. This was not mainstream science so much as conservative, lowest-common-denominator science. But when the IPCC finally announced its cautious conclusions, they had solid credibility.

Issued in 1990, the first IPCC report concluded that the world had indeed been warming. Much of this might possibly be attributed to natural processes, the report conceded. It predicted (correctly) that it would take another decade before scientists could be confident that the change was caused by the greenhouse effect. But the panel thought it likely that there could be a warming of several degrees by the year 2050. There was nothing exciting or surprising in the report. It barely qualified as "news" and got hardly any newspaper coverage.

A closer look at the IPCC report would have found more food for thought. The scientists had tried to draw attention to other greenhouse gases, less abundant than CO_2 but nevertheless potent. Controlling these might offer an economically sound way to get a start on reducing the risk of warming. When the panel said the future was uncertain, they did not mean that the potential for harm should be ignored.

Other groups, ranging from government agencies to environmentalist organizations, had begun to suggest a variety of steps that might be taken. A 1991 National Academy of Sciences report listed no less than 58 policies that could mitigate greenhouse warming. Some were "no-regrets" policies, so practical that they would benefit the economy whether or not there was a global warming problem. For example, governments might promote improvements in the efficiency of commercial lighting, home heating, and trucks. Or they could reduce the costly subsidies that encouraged wasteful consumption of fossil fuels. Some policies would carry a modest cost that would be compensated by valuable social benefits. Why not, for example, devise ways to reduce car commuting time, and reforest overgrazed wastelands? Some ideas were too expensive at present, but might become practical if technology was driven forward by the regulation or taxation of greenhouse gas emissions, or by plain desperation. For example, it might someday make sense to extract CO_2 as a power plant burned fuel, and sequester the gas in the depths of the oceans or underground. And some proposals were visionary. Couldn't we replace fossil fuels by growing crops that stored energy from sunlight, or launch flotillas of mirrors into orbit to reflect sunlight away from the Earth?

None of this attracted much public interest. After the flood of global warming stories in 1988, media attention had inevitably declined as more normal weather set in and editors looked for newer topics. Environmentalist organizations carried on with sporadic lobbying and advertising efforts to warn about global warming and argue for restrictions on emissions. They were opposed, and greatly outspent, by industries that produced or relied on fossil fuels. Industry groups not only mounted a sustained and professional public relations effort, but also channeled considerable sums of money to individual scientists and small conservative institutions and publications that denied any need to act against global warming. This effort followed the pattern of scientific criticism and advertising that industrial groups had used to attack warnings against ozone depletion and acid rain. Although those campaigns had been discredited after a decade or two, fair-minded people were ready to listen to the global-warming skeptics.

It was reasonable to argue that intrusive government regulation to reduce CO_2 emissions would be premature, given the scientific uncertainties. Conservatives pointed out that if something did have to be done, the longer we waited, the better we might know how to do it. They also argued that a strong economy (which they presumed meant one with the least possible government regulation of industry) would offer the best insurance against future shocks. Activists replied that action to retard the damage should begin as soon as possible, if only to gain experience in how to restrict gases without harming the economy. They argued hardest for policy changes that they had long desired for other reasons, such as protecting tropical forests and removing government subsidies that promoted fossil fuel use.

The main argument against taking action was to deny that warming was likely at all. Critics cited valid grounds for scientific skepticism, which the citizen with a taste for science could pick up from occasional semi-popular articles. Even the tentative statements in the IPCC's consensus report could not represent the particular views of every scientist on such a complex and uncertain problem. Surveys of climate experts in the early 1990s found that most of them felt their understanding of climate change was poor, and predictions remained highly uncertain—even more uncertain than indicated by the IPCC's report (at least as the news media described it). A majority of climate experts did believe that significant global warming was likely to happen, even if they couldn't prove it. Asked to rank their certainty about this on a scale from one to ten, the majority picked a number near the middle. Only a few climate experts were fairly confident that there would be no global warming at all, although as they pointed out, scientific truth is not reached by taking a vote. Roughly two-thirds of the scientists polled felt that there was enough evidence in hand to make it reasonable for the world to start taking policy steps to lessen the danger, just in case. On one thing nearly all scientists agreed: the future was likely to hold "surprises," deviations from the climate as currently understood.[1]

For those who wanted to convince the public that greenhouse

warming was a chimera, the first issue to confront was the global temperature itself. News media now reported fairly prominently the annual summaries of the planet's temperature issued by the New York and East Anglia groups (joined by a NOAA Data Center in Asheville, North Carolina). The year 1988 proved to be a record-breaker. But in the early 1990s, average global temperatures dipped.

A few scientists claimed that there had been no real global warming trend at all, that the statistics of record-breaking heat since the 1970s were illusory. The most prominent of these skeptics was S. Fred Singer, who had retired in 1989 from a distinguished career managing government programs in weather satellites and other technical enterprises, then founded an environmental policy group funded by conservative foundations. Singer argued, for example, that all the expert groups had somehow failed to properly account for the well-known effects of urbanization when they compiled weather statistics. Other skeptics conceded that global temperatures had risen modestly, but held that the rise was just a chance fluctuation. After all, gradual drops and rises of average temperature around the North Atlantic in particular had been seen for centuries. Why couldn't the next decades experience a cooling? They entirely disbelieved the computer models that predicted warming from the greenhouse effect.

Leading modelers admitted they did have problems. For example, their models typically had a strong "sensitivity" to greenhouse gases, predicting roughly 3°C of global warming for a doubling of CO_2. As a few critics pointed out, for the known increase of greenhouse gases over the twentieth century, the models calculated a one-degree rise, but the temperatures actually recorded had risen at most half a degree. Another galling problem was that different models gave different predictions for just how global warming would affect a given locality . That was useless for the senator who wanted to know whether the greenhouse effect would make his state wetter or dryer in the next century. In 1992, a massive study was published comparing 14 General Circulation Models constructed by groups in eight different nations. As usual, the GCMs differed in many details, and in a few respects they all erred in the same direction. For example, they all got the present tropics a bit too cold. Perhaps that reflected some flaw all the models shared, something they had collectively overlooked. If so, was the flaw so bad that it made their predictions altogether worthless?

Nevertheless, most experts felt the GCMs were on the right track. In the 14-model comparison, all the results were at least in rough overall agreement with present reality. Feeding an increase of CO_2 into the different models produced more or less the same temperature trend—not in any given region, perhaps, but averaged over the globe. It seemed impossible to construct a realistic climate model that did not show some kind of greenhouse-effect warming.

A substantial number of scientists continued to doubt that the models could be trusted at all. The technical criticism most widely noted in the press came in several brief "reports"—not scientific papers in the usual sense—published between 1989 and 1992 by the conservative George C. Marshall Institute. The publications came with the endorsement of Frederick Seitz, a former head of the National Academy of Sciences, highly admired for his past achievements in administration and the physics of solid materials. The pamphlets themselves were anonymously authored, but they assembled a well-argued array of skeptical scientific thinking, backed up by vocal support from a few reputable meteorologists. Concerned that proposed government regulation would be "extraordinarily costly to the U.S. economy," the reports insisted it would be unwise to act on the basis of the feeble global warming theories.[2]

Scientists noticed something that the public largely overlooked: the most outspoken scientific critiques of global warming predictions rarely appeared in the standard scientific publications, the "peer-reviewed" journals where every statement was reviewed by other scientists before publication. With a few exceptions, the critiques tended to appear in venues funded by industrial groups and conservative foundations, or in business-oriented media like the *Wall Street Journal*. Most climate experts, while agreeing that future warming was not a proven fact, found the critics' counterarguments dubious. Some publicly decried their reports as misleading "junk science."[3] On several points open conflict broke out among the scientists, with acrimonious and personalized exchanges.

To science journalists and their editors the controversy was confusing, but excellent story material. A study of American media covering the years 1985 to 1991 found substantial coverage of climate change in newspapers and news magazines, although only occasional stories on television. Outside the most deeply conservative media, reporters tended to accept the idea that greenhouse warming was under way. Following the usual tendency of the media to grab attention with dire predictions, a majority of the reports played up cataclysmic possibilities: devastating droughts, ferocious storms, waves attacking drowned coastlines.[4] Emphasizing conflict, as was their wont in covering almost any issue, some reporters wrote their stories as if the issue were a simple fight between climate scientists and the Republican administration. Many other reports presented the issue as if it were a quarrel between two diametrically opposed groups of scientists. Newspaper and magazine articles and television reports often sought an artificial balance by pitting "pro" against "anti" scientists, one to one. It was hard to recall that there was a consensus view, shared by most experts: the IPCC's modest conclusion that global warming was probable although not certain. The media did get one thing right in the early 1990s, the fact that most scientists emphasized the lack of certainty.

The criticism of global-warming predictions strongly influenced President George H. W. Bush's administration, inclined as it was to agree with the sort of views propounded by columnists in the *Wall Street Journal*. Some federal agencies were in favor of action against greenhouse gases, notably the Department of Energy, the Environmental Protection Agency, and the Department of State (under pressure from European governments). But many others in the administration, as in the Reagan administration, only wished the issue would somehow disappear. A White House memorandum, inadvertently released in 1990, proposed that the best way to deal with concern about global warming would be "to raise the many uncertainties."[5]

Uncertainty was easy to raise when the skepticism of a few energetic scientists was widely circulated in publications sponsored by conservative groups and industrial interests. In the forefront was the Global Climate Coalition, generously funded by dozens of major corporations in the petroleum, automotive, and other industries. With slick publications and videos sent wholesale to journalists, plus extensive personal lobbying in Washington and at international meetings, the Coalition did much to persuade leaders who were ignorant of science that there was no sound reason to worry about climate change. Enough of the public was likewise sufficiently persuaded by the skeptical advertising and news reports, or at least sufficiently confused by them, so that the administration felt free to avoid taking serious steps against global warming.

The U.S. government's stubborn rejection of the IPCC report became an embarrassment in 1992. World leaders were preparing their grandest meeting ever, an "Earth Summit" in Rio de Janeiro (officially, the United Nations Conference on Environment and Development). In other advanced industrial nations, a combination of scientific advice and political pressure from Green political parties and the environmentalist public was raising deep official concern about global warming. The great majority of countries, led by the Western Europeans, called for mandatory limits on greenhouse gas emissions. But no agreement could get far without the United States, the world's premier political, economic, and scientific power—and largest emitter of greenhouse gases.

The American administration, attacked by its closest foreign friends as an irresponsible polluter, showed some flexibility. The negotiators papered over disagreements to produce a compromise that included targets for reduced emissions. An agreement (officially the "United Nations Framework Convention on Climate Change") was signed at Rio by more than 150 states. The agreement's evasions and ambiguities left governments enough loopholes so they could avoid meaningful action. But it did establish some basic principles for what governments ought to do eventually, and it pointed out a path for further negotiation. The Bush administration followed up with a number of inexpensive "no regrets" policies to promote energy efficiency. These were far too modest to meet the targets for reduced emissions. Most other nations felt little pressure to try to do much more.

The creation of the IPCC had established a cyclic international process. Roughly twice a decade, the IPCC would gather together the most recent research and issue a consensus statement about the prospects for climate change. That would lay a foundation for international negotiations, which would in turn give guidelines for individual national policies. Further moves would await the results of further research. In short, after governments responded to the Rio convention, it was the scientists' turn. Although scientists pursued research problems as usual, published the results for their peers as usual, and discussed the technical points in meetings as usual, to officialdom this was all in preparation for the next IPCC report, scheduled for 1995.

The computer modeling work pushed ahead without any big breakthroughs, a steady, slogging job. One group would find a better way to represent the formation of clouds, another would come up with a more efficient way to calculate winds, and once or twice a decade each group would convince some agency to buy them a computer many times faster than the last. Thanks to the fabulous expansion of processing power, groups could now routinely couple together ocean and atmosphere models to represent the climate system as a whole in elaborate detail. Their work was greatly helped by the fact that they could set their results against a uniform body of worldwide data. Specially designed satellite instruments were monitoring incoming and outgoing radiation, cloud cover, and other essential parameters—showing, for example, where clouds brought warming and where they brought cooling.

The most significant scientific development was a growing recognition that global climate change was not a matter of CO_2 alone. Besides other greenhouse gases like methane, human activity was providing at least a quarter of the atmosphere's dust, chemical haze, and various other aerosol particles. All these agents had a variety of effects on incoming and outgoing radiation, directly or through their influence on clouds. In particular, as workers in various fields ground out measurements and calculations and exchanged information, they began to agree that sulfate aerosols brought significant cooling.

In 1991 Mount Pinatubo in the Philippines exploded. A mushroom cloud the size of Iowa burst into the stratosphere, depositing some 20 million tons of SO_2. Hansen's group saw an opportunity in this "natural experiment," the twentieth century's biggest volcanic injection of aerosols into the stratosphere. It could provide a strict test of their computer model. From their calculations they boldly predicted roughly half a degree of average global cooling. It would be concentrated in the higher northern latitudes and would last a couple of years. Exactly such a temporary cooling was in fact observed—this was the pause in record-setting heat waves that encouraged global warming skeptics. The success of Hansen's prediction strengthened the modelers' confidence that they could get a grip on the effects of aerosols.

As one expert dryly remarked, "the fact that aerosols have been ignored means that projections may well be grossly in error."[6] The "human volcano" was steadily emitting a large amount of sulfates and other aerosols. Apparently this acted like an ongoing Pinatubo eruption, offsetting some of the greenhouse-effect warming. Computer modelers set to work to incorporate the improved understanding of aerosols into their GCMs. Their results refuted the scientists who had criticized the models for false "sensitivity"—meaning that when the known increase of CO_2 in the twentieth century was put in, models calculated a rise of temperature double what had actually been observed. It turned out that the models had calculated the effects of CO_2 well enough. But now that modelers were at last able to figure in the effects of increased aerosol pollution, the calculated cooling did partly counteract the expected warming. In 1995, improved GCMs constructed at three different centers (the Lawrence Livermore National Laboratory in California, the Hadley Center for Climate Prediction and Research in the United Kingdom, and the Max Planck Institute for Meteorology in Germany) all reproduced the general trend of twentieth-century temperature changes.

There was further convincing support for greenhouse-effect predictions. More than a century back, Arrhenius had figured on elementary principles that the effect would act most strongly at night (when the planet loses heat most rapidly into space). Statistics did show that it was especially at night that the world was warmer. Moreover, Arrhenius and everyone since had calculated that the Arctic would warm more than other parts of the globe (as the melting of snow and ice exposed dark soil and water). The effect was glaringly obvious to scientists as they saw mountain meadows taken over by trees in Sweden and the Arctic Ocean icepack growing thin. Alaskans and Siberians didn't need statistics to tell them the weather was changing when they saw buildings sag as the permafrost that supported them melted.

Pursuing this in a more sophisticated way, computer models predicted that greenhouse gases would cause a particular regional pattern of temperature change. It was different from what might be caused by other external influences (for example, solar variations). The observed geographical pattern did bear a rough resemblance to the GCMs' greenhouse-effect maps. This was a long-sought "fingerprint" of the greenhouse effect, strong evidence that it was happening as the models predicted. These results exerted a powerful influence when the IPCC went to work on its next report. In numerous workshops and exchanges of papers, experts pored over a great variety of evidence and calculations. But what impressed them most was the way GCMs, now that they incorporated aerosol pollution along with so much else, showed a convincing detailed correspondence with the century's actual global temperature trend.

After another grueling round of analysis, negotiation, and lobbying among some 400 scientist experts plus representatives of every variety of national and nongovernmental interests, in 1995 the IPCC gave the world its conclusions. The report's single widely quoted sentence said, "The balance of evidence suggests that there is a discernible human influence on global climate." The weasely wording showed the strain of political compromises that had watered down the original draft, but the message was unmistakable. "It's official," as *Science* magazine put it—the "first glimmer of greenhouse warming" had been seen.[7]

This second IPCC report estimated that a doubled CO_2 level—which was expected sometime around the mid-twenty-first century—would raise the average temperature somewhere between 1.5 and 4.5°C. That was exactly the range of numbers announced in the first IPCC report and by other groups ever since 1979, when the Charney Committee of the U.S. National Academy of Sciences had published 3°C, plus or minus 1.5°C, as a plausible estimate (Chapter 5). Since then computer modeling had made tremendous progress. The latest scenarios actually suggested a somewhat different range of possibilities, rising up to 5.5°C or so. But the meaning of these numbers had been hazy from the beginning. All they represented was what a group of experts found reasonable. The scientists who wrote the 1995 IPCC report decided to stick with the Charney Committee's familiar numbers, rather than give critics an opening to cry inconsistency. The meaning of the numbers had invisibly changed: the experts had grown more confident that the warming would in fact fall within this range. It was a striking demonstration of how the IPCC process deliberately mingled science and politics until they could scarcely be disentangled.

The somnolent public debate revived in late 1995 on the news that the IPCC had agreed that the world was indeed getting warmer, and that the warming was probably caused at least in part by humanity.[8] Although many scientists had been saying as much for years, this was the first formal declaration by the assembled experts of the world. It was page-one news everywhere, immediately recognized as a landmark. Better still for journalists, the report stirred up a nasty controversy, for a few critics cast doubt upon the personal integrity of some IPCC scientists.

Most national governments were more willing than before to respond. In the United States, after Bill Clinton took office as President in 1993, Vice President Al Gore and others persuaded him to commit the nation formally to the Rio target for reducing greenhouse gases. But conservatives retained a predominant influence in Washington's politics. Many powerful conservatives not only scoffed at any research that pointed to environmental problems, but held deep suspicions about the United Nations and all its cooperative international programs. Faced with these ardent opponents, Clinton was unwilling to spend his limited political capital on an issue that would not become acute during his term in office.

Neither criticism nor official indifference stopped the international process from pushing forward according to its negotiated schedule. The agreements worked out at Rio led to a new conclave, a 1997 U.N. Conference on Climate Change held in Kyoto, Japan. It was a policy-and-media extravaganza attended by nearly 6,000 official delegates and thousands more representatives of environmental groups and industry, plus a swarm of reporters. Representatives of the United States proposed that industrial countries gradually reduce their emissions to 1990 levels. Most other governments, with Western European countries in the lead, demanded more aggressive action. Coal-rich China and most other developing countries demanded exemption from regulation, however, until their economies caught up with the nations that had already industrialized. The greenhouse debate had now become tangled up with intractable problems involving fairness and the power relations between industrialized and developing countries. As a further impediment, the groups with the most to lose from global warming—poor people and generations unborn—had the least power to force through an agreement. The negotiations almost broke down in frustration and exhaustion. But the IPCC's conclusions could not be brushed aside. Dedicated efforts by many leaders were capped by a dramatic intervention by U.S. Vice President Al Gore, who flew to Kyoto on the last day and pushed through a compromise—the Kyoto Protocol. The agreement exempted poor countries for the time being, and pledged wealthy countries to cut their emissions significantly by 2010.

According to the agreement, national governments were supposed to incorporate their pledges into concrete policies. To block any chance of that, in the United States the Global Climate Coalition of major industrial corporations mounted a multimillion-dollar lobbying and advertising campaign. The effort was aided by the small but respectable set of scientists who continued to question the temperature statistics and computer models, but the chief arguments were political. Conservatives exclaimed that regulation would inflict economic disaster on Americans, and appealed to nationalism by warning that the Kyoto Protocol would turn over the world economy to the unregulated developing countries. They pointed with horror to the specter of a "carbon tax," a levy on CO_2 emissions that would raise gasoline prices—something Americans supposedly could never tolerate (unlike, say, Europeans). In the polarized debates, scarcely anyone remarked that more subtle approaches to averting greenhouse effect warming were possible—methods that might be both effective and politically acceptable.

Even before the Kyoto delegates had assembled, the U.S. Senate had declared by a vote of 95–0 that it would reject a treaty that failed to set limits for developing countries. Afterwards, the treaty was never submitted to the Senate for ratification. With little debate, Congress declined to make any policy changes that might help move toward meeting the Kyoto targets. That gave other nations an excuse to continue business as usual. Yet even if the Kyoto agreement had been taken up more aggressively, people on both sides of the debate pointed out that it would have made only a start. It embodied so many compromises, and so many untested mechanisms for setting standards and enforcement, that the agreement could scarcely force a stabilization of emissions, let alone a reduction.

So it was again the turn of the scientists. A few skeptics continued to offer a variety of plausible technical criticisms, stimulating research in many directions. A particularly influential critique, widely publicized in a Marshall Institute report, argued that the most probable cause of the twentieth century's global warming (if it existed at all) was just a temporary increase in solar activity.

In truth, something besides CO_2 and aerosols was needed to make an exact match with the peculiar rise-fall-rise of temperature since the nineteenth century. Sunspots and other measures plainly showed that the Sun's activity had followed a similar trend. A solar connection such as the one that Jack Eddy had sketched out sounded increasingly plausible. For example, an analysis of satellite observations in the 1990s found that global cloudiness had increased slightly at times of weaker solar activity. A few scientists proposed mechanisms for such an effect, complex processes involving the effects of cosmic rays or ultraviolet radiation on water-droplet formation or on ozone. Experiments with GCMs suggested that even such tiny variations might indeed make a difference, by interfering in the teetering feedback cycles that linked stratospheric chemistry and particles with lower-level winds. Whatever the exact mechanism, most scientists came to accept that the climate system was so unsteady that minor changes of the Sun's radiation might provoke significant shifts. Most experts now thought it likely that the warming trend from the 1880s to the 1940s, when greenhouse emissions were not yet great, had been caused at least in part by increased solar activity.

Some rough limits could be set on the extent of the Sun's influence. The average solar activity had not increased further during the 1980s and 1990s. Yet the global temperature rise that began in the 1970s had continued through the 1980s. In the 1990s, once the aerosols from the Pinatubo eruption washed out of the atmosphere, the rise accelerated. It was getting hard to explain this without invoking greenhouse gases. The significance of the argument that solar variations influenced climate was now reversed. If the planet reacted with such extreme sensitivity to minor changes in the radiation arriving from the sun, it had to be sensitive to greenhouse-gas interference with the radiation once it entered the atmosphere. A 1994 National Academy of Sciences panel estimated that if solar radiation were now to weaken as much as it had in the Little Ice Age of the seventeenth century, the effect would be offset by only two decades of accumulation of greenhouse gases. As one expert explained, the Little Ice Age "was a mere 'blip' compared with expected future climatic change."[8]

Such predictions of warming depended entirely on computer models. Here the critics still found strong reasons for doubt. While the best GCMs could now reproduce the present climate, that was far from guaranteeing they could reliably predict a quite different future climate. The models got the present climate right only because they had been laboriously "tuned" to match it, by adjusting a variety of arbitrary parameters. A few scientists who remained skeptical brought this problem into public view, accusing the modelers of fudging their results. But the arguments were too technical to attract much attention, and most of the debate was carried out within scientific journals and conferences.

Thus, for example, the respected Massachusetts Institute of Technology meteorologist Richard Lindzen challenged the way modelers allowed for water-vapor feedback. This was the crucial calculation showing how a warmer atmosphere would carry more water vapor, which would in turn amplify any greenhouse effect. Lindzen offered an alternative scenario involving changes in the way moisture was carried up and down between layers of the atmosphere. While Lindzen's detailed argument was complex, he said his thinking rested on a simple philosophical conviction: over the long run natural self-regulation must always win out. Few scientists found Lindzen's technical argument convincing (and his credibility was not helped when he accepted money as a consultant for fossil-fuel groups). The bulk of evidence indicated that the way the modelers handled water vapor, although far from perfect, was not wildly astray.

Another criticism from insiders was that models had attempted a crucial test—matching an actual different climate—and failed. Back in 1976, oceanographers had cooperated in a large-scale project, CLIMAP, trolling the seven seas for information on conditions at the peak of the last ice age. From measurements of foraminifera shells brought up from the seabed, they created a global map of ocean temperatures roughly 20,000 years ago. The CLIMAP team reported that in the middle of the last ice age, tropical seas had been only slightly cooler than at present. That raised doubts about whether the climate really was as sensitive as the modelers thought to external forces like greenhouse gases. Moreover, while the seas had stayed warm during the last ice age, the air at high elevations had been far colder. That was evident, for example, in lower altitudes of former snowlines detected on the mountains of New

Guinea and Hawaii. And no matter how much the GCMs were fiddled, they could not be persuaded to show such a large difference of temperature with altitude. Were they fundamentally flawed?

In the late 1990s the problem was largely resolved. It turned out that what had been unreliable was not the computer models, but the oceanographers' complex manipulation of their data. A variety of new types of climate measures indicated that tropical ice-age waters had turned significantly colder, perhaps 3°C or more, agreeing reasonably well with the GCMs. Debate continued, for various types of climate records disagreed. But nearly all scientists were now confident that the computer models rendered actual climate processes faithfully enough to make reliable predictions.

Many problems that the models still could not confront remained, and it was necessary to work around them with crude approximations. For example, modelers used average parameters to derive the effects of cloud cover, unable to calculate from first principles the interactions among radiation, water droplets, and aerosols. As the twenty-first century began, experts were still speculating about a variety of subtle mechanisms involving changes in clouds, which might significantly affect the models' predictions. Even if they knew exactly how particles affected clouds and radiation, modelers would have to know what particles would be present in the first place. That meant learning much more about the chemistry of the atmosphere, where particles were brewed up by a variety of pollutants—which were mostly measured sketchily and changed from year to year. Worse, oceanographers still had not solved the mystery of how heat is transported up or down from layer to layer in the seas. Until the real processes could be observed and represented with equations, the risk remained that the ocean circulation could change in some way radically different from what the computers calculated. Finally, even a perfect atmospheric GCM, perfectly coupled to a perfect ocean circulation model, would only make a start. For one thing, it would have to be coupled to models of vegetation.

By the mid 1990s, scientists had found convincing evidence that changes in vegetation could alter regional climates. At several locations, overgrazed grasslands with dried-out soils had become demonstrably hotter than less used pastures (and the heating would make it all the harder for grass to return). Some rain forests that were deforested showed a measurable decrease in rainfall, since moisture was no longer evaporated back into the air from the leaves of trees—in Brazil, rain fled from the plough. It was also pointed out that if global warming made forests grow farther north, the dark pines would absorb more sunlight than snowy tundra, heat the air, and add to global warming. In whatever ways vegetation was altered, whether through direct human action or by a shift of climate, there could be strong feedbacks with a potential for a lasting, self-sustained change.

Some scientists stuck by the old view that biological feedbacks were reassuring rather than alarming. They held that fertilization from the increased CO_2 in the atmosphere would benefit agriculture and forestry so much that it would make up for any possible damage from climate change. Studies found that for the planet as a whole, biomass was indeed absorbing more CO_2 overall than in earlier decades. However, the consequences were not straightforward. For example, under some circumstances the extra CO_2 might benefit weeds and insect pests more than desirable crops. In any case, as the level of the gas continued to rise, plants would reach a point (nobody could predict how soon) where they would be unable to use more carbon fertilizer. There was a good chance that more warmth would eventually foster decay, with a net *emission* of greenhouse gases. Meanwhile new evidence showed that the biology of the oceans too could vary markedly. The drifting plankton, as complex as a rain forest but scarcely explored by biologists, interacted strongly with the uptake or emission of CO_2. All this had to be incorporated in models.

The whole global system was so complex that the old original puzzle—how had the ice ages started and ended?—stood unsolved. Feeble changes in sunlight were perhaps only the last straw to tilt a balance involving the dynamics of the growth and melting of massive ice sheets, and also changes in vegetation and the bio-geochemistry of the oceans, all interacting with greenhouse gases. When people talked now of a GCM they no longer meant a General Circulation Model, built from the traditional equations for weather. GCM now stood for Global Climate Model or even Global Coupled Model, incorporating many things besides the circulation of the atmosphere.

Although the modelers knew little about many of these factors, they did seem to be taking all the most important forces into account somehow, once they had adjusted enough parameters in the models. For the models did agree on reasonable results for such different conditions as arctic and tropics, ocean and desert, winter and summer. They could now reproduce pretty well the picture of Earth after a major volcanic eruption and even during an ice age. All this raised confidence that climate models could not be too far wrong in their disturbing predictions of future change. The possibility remained that the models all shared some hidden flaw, but if so, it was not likely to show up until greenhouse gases pushed the climate beyond any conditions that the modelers had attempted to calculate. The possibility of flaws in the models did not mean, as some critics implied, that people did not need to worry about global warming. As Broecker and many others pointed out, if the models were faulty, future climate change could be *worse* than predicted, not better. The interactions among winds, oceans, ice sheets, forests, and so forth might be so unstable that scientists would never be able to predict a future climate for sure, any more than meteorologists could predict a day of rain a year in advance.

Still, when it came to the modest but vital problem of predicting the climate of the next century or so, by the start of the twenty-first century the modelers could confidently declare what was *reasonably likely* to happen. The protracted research efforts of a dozen teams had converged on answers. Even the most prominent scientist critics quietly admitted that sooner or later the greenhouse effect must be felt. The old predictions were solid. Doubling the CO_2 level was almost certain to raise the average temperature, and most likely by around 3°C, give or take a degree or so.

Different GCMs stubbornly continued to give different predictions for particular regions. Some things looked quite certain, like higher temperatures in the Arctic (hardly a prediction now, for such warming was blatantly visible). But for many of the Earth's populated places, the models could not reliably tell the local government whether to brace itself for droughts, floods, or neither or both. Yet with more heat and moisture driving the engines of weather, it was likely that the world would see stronger storms, greater floods, and worse droughts. Still more certain was the way the heat already in the atmosphere must work its way down into the ocean depths. Since water expands when heated, it was hard to deny the sea level will rise. By the late twenty-first century, coastal areas from New Orleans to Bangladesh will suffer serious everyday difficulties and occasional calamitous storm surges. Low-lying areas where tens of millions of people now live are likely to be obliterated. In later centuries it will get worse, for even if global warming is halted, the heat will continue to work its way gradually deeper into the oceans and the tides will continue to creep higher.

Most politicians barely noticed these problems among the many urgent demands for their attention. They would not go against short-term industrial interests, unless public opinion drove them. But debates over global warming, like most political controversies, did not mobilize the media for long periods. In between episodes of controversy, the issue slipped away from public attention. Politicians saw little to gain by stirring it up. Even Gore mentioned global warming only briefly during his 2000 run for the presidency.

The world's image makers had failed to give the public a vivid picture of what climate change might truly mean. There was nothing like the response to the threat of nuclear war in earlier decades, when first-class novels and movies had commanded everyone's attention. Global warming featured in a bare handful of science fiction paperbacks and shoddy movies, where scientifically dubious monster storms or radical sea-level rise served as a background for hackneyed action plots. The general public was never offered convincing and humanized tales of travails that might realistically beset us: the squalid ruin of the world's mountain meadows and coral reefs, the mounting impoverishment due to crop failures, the invasions of tropical diseases, the press of millions of refugees from drowned coastal regions.

Occasionally, science reporters would find a news hook for a story. The press took mild notice, for example, when the groups compiling statistics announced that 1995 was the warmest year on record for the planet as a whole, and when 1997 broke that record, and 1998 yet again. The impact was muted, however, for the warming was most pronounced in remote ocean and arctic regions. Some smaller but important places—in particular the U.S. East Coast, with its key political and media centers—had not experienced the warming that was evident in some other regions in the latter decades of the century.

Reports of official studies each had their day in the limelight, but rarely more than a day. A story made more of an impression if it dealt with something visible, as when ice floes the size of a small nation split off from Antarctica. Other chances to mention global climate change came in reports on heat waves, floods, and coastal storms, especially when the events were more damaging than anything in recent memory. In fact, any one of these widely reported incidents might have had nothing to do with global warming. But they symbolically conveyed what scientists did believe. By the late 1990s, there were many kinds of valid indicators of global warming. For example, the Northern Hemisphere spring was coming on average a week earlier than in the 1970s.

A variety of new evidence suggested that the recent warming was exceptional even if one looked back many centuries. Old temperatures could be estimated from historical records of events like freezes and harvests, or derived from analysis of tree rings, coral reefs, and so forth. An example of the far-flung efforts was a series of heroic expeditions that labored high in the thin air of the Andes and Tibet to drill into tropical ice caps. These too showed that the warming in the last few decades was above anything seen for thousands of years before. Indeed, the ice caps themselves, which had endured since the last ice age, were melting away faster than the scientists could measure them. A widely reprinted graph that compiled estimated temperatures over the past ten centuries showed a sharp turn upward since the start of the Industrial Revolution and especially in the most recent years. Apparently 1998 had been not just the warmest year of the century, but of the millennium.

The public barely noticed the revelations that most startled climate scientists in the 1990s. The first shock came from the center of the Greenland ice plateau. Early hopes for a new cooperative program between Americans and Europeans had broken down, and each team set to drilling its own hole. Competition was transmuted into cooperation by a decision to put the two boreholes just far enough apart so that anything that showed up in both cores must represent a real climate effect, not an accident due to bedrock conditions. The match turned out to be remarkably exact for most of the way down. The comparison between cores showed, convincingly, that climate could change more rapidly than almost any scientist had imagined.

Swings of temperature that were believed in the 1960s to take tens of thousands of years, in the 1970s to take thousands of years,

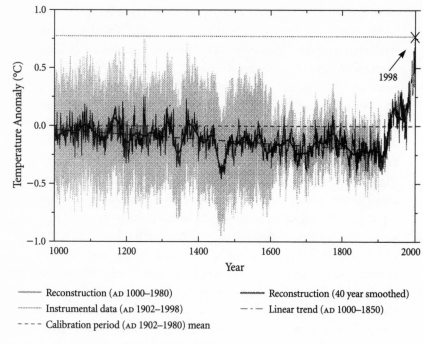

Figure 3. UNPRECENDENTED RECENT RISE OF GLOBAL
TEMPERATURE.

The "hockey stick" graph: a reconstruction of temperatures averaged over the globe for the past millennium, plus measured temperatures for the past century. A dashed line tracks hints of a downward trend, interrupted in recent decades just when greenhouse gases soared. (M. Mann et al., *Geophysical Research Letters* 26, p. 761, 1999, copyright 1999 American Geophysical Union, reproduced by permission.)

and in the 1980s to take hundreds of years, now were discovered to take only decades. During the last glacial period, Greenland had sometimes warmed as much as 7°C in the space of less than 50 years. During the Younger Dryas transition, spectacular shifts in the entire North Atlantic climate were visible within only five snow layers, that is, five years! The evidence could no longer be dismissed as incredible, for at least one explanation was at hand. Computer models showed the possibility of a rapid and drastic flip-flop of the North Atlantic circulation. Meanwhile various kinds of geological evidence from other continents suggested that the Younger Dryas had brought climate changes not only around the North Atlantic, but around the globe.

Could such variations occur not only in glacial times, but also in a warm interglacial period like our own era? Computer models and other evidence showed it could happen—indeed, we might cause it. Paradoxically, global warming might bring devastating regional cooling across a stretch of the planet from Chicago to Moscow. "There is surely a possibility," warned Broecker, "that the ongoing buildup of greenhouse gases might trigger yet another of these ocean reorganizations . . . it could lead to widespread starvation."[9] Other mechanisms for catastrophic transformations remained on the table. For one thing, adventurous surveys across the West Antarctic Ice Sheet confirmed the dismaying possibility that it might collapse within the next few centuries. For another, new studies found it plausible that warming of the oceans could trigger a disintegration of deposits of clathrate ices within the seabed mud. That could vent enough methane and CO_2 into the atmosphere to bring immense warming. Geological evidence suggested that such a catastrophe had really happened at least once, 55 million years back, causing vast extinctions. Most ominous of all, in 1993 Dansgaard and his colleagues reported that their Greenland cores showed horrendous oscillations during the last interglacial warm period—for example, a 14° cold snap that had struck in the space of a decade.

This last item turned out to be an illusion. The measurements, made in ice pulled up from near bedrock, had been distorted by glacial flow that stirred together layers from warm and cold periods. But now that scientists had been forced to think in a new way about the climate system, they did not look back. The undoubted temperature jumps seen in ice cores were rapid and severe enough.

People also recalled that there was plenty of evidence for a sudden onset of prolonged droughts, like the one that had devastated North American native cultures in the 1200s (Chapter 4). New geological evidence pointed to such droughts at the time of the downfall of Mayan and ancient Mesopotamian cultures too. The standard picture of benign, equable interglacials was gone beyond retrieval.

The most disturbing evidence came from the ice core drilled at Vostok in Antarctica. In this record reaching back over nearly four complete glacial-interglacial cycles, almost every stretch was peppered with drastic temperature changes. When Bryson, Schneider, and others had warned that the century or so of stability in recent memory did not reflect "normal" long-term variations such as the Little Ice Age (Chapter 5), they had touched on an instability grander than they guessed. The entire rise of human civilization since the end of the Younger Dryas had taken place during a warm period that was far more stable than any other period in the last 400,000 years. The climate known to history was a lucky anomaly.

Despite the profound implications of this new viewpoint, hardly anyone rose to dispute it. It became part of the consensus of nearly all the world's climate scientists, as embodied in the IPCC's 1995 report. The report included a notice that climate "surprises" were possible—"Future unexpected, large and rapid climate system changes (as have occurred in the past)."[10] But the point was not emphasized by the authors, and it was seldom mentioned in the press. To everyone but the climate experts, and even to many of them, future "climate change" meant the gradual warming that was now becoming palpable in many parts of the world.

Getting a firmer grasp on the possibilities called for a new round of the IPCC process. Again scientists gathered in groups to sort through and debate all the newest scientific results. In the negotiations that laid out the panel's third report, issued in 2001, the consensus of scientists overwhelmed objections from industry-oriented skeptics. The IPCC bluntly concluded that the world was rapidly getting warmer. And strong new evidence showed that "most of the observed warming over the last 50 years is likely to have been due to the increase in greenhouse gas concentrations." Moreover, computer modeling had improved to the point where the panel could conclude, more confidently than ever, that global temperatures would rise much higher still. Indeed, the rate of warming was "very likely to be without precedent during at least the last 10,000 years." Under a worst-case scenario, where global emissions of CO_2 and restrictions on sulfate pollution might rise faster than previous reports had considered, the range of warming predicted for the late twenty-first century ran from 1.4°C up to a shocking 5.8°C (10°F).[11] (The panel again mentioned the possibility of still more abrupt climate surprises, and again that was mostly overlooked.)

This predicted temperature range was not for the traditional doubled CO_2 level, expected around mid-century, but for the still higher levels expected after 2070. As Broecker pointed out, "While previously we thought in terms of doubling the strength of the CO_2 content of the preindustrial atmosphere, current thought is moving toward a tripling." Eventually the level would move higher still, if not halted by self-restraint or catastrophe.[12]

The IPCC report haunted the next great international conference, held at The Hague in late 2000. Although the report was not yet completed, its main conclusions had been leaked to the delegates. Representatives from 170 countries assembled to write the specific rules that might force reductions in greenhouse gases as promised at Kyoto. Continental Europeans insisted on a strict regime of greenhouse gas regulation. That approach found no effective political backing in the United States, whose government insisted on more market-friendly mechanisms. The negotiations collapsed. The final blow came in March 2001, when the newly installed American President, George W. Bush, rejected any significant regulation of the nation's CO_2 emissions, publicly renouncing the Kyoto protocols.

International diplomacy is a gradual process. The truly important work is to shift attitudes step by step, and at the same time to create mechanisms (such as ways to measure national emissions and processes to adjudicate quotas)—mechanisms which might be hollow at the start but could slowly become meaningful. The people who denied any need for action on global warming were becoming isolated and left behind. The first major business group to get worried had been the insurance industry. In the early 1990s it endured mammoth losses as storms and floods increased, just as Hansen and other global-warming theorists had predicted. By the end of the 1990s, a number of other key companies had acknowledged that greenhouse warming was a real problem for them, and they quit the Global Climate Coalition. Some began to restructure their operations so that they could flourish in a warming world with restrictions on emissions. The *Economist* magazine, a free-market champion, reported that corporations now wanted "clear ground-rules for the green energy projects, clean-development schemes and emissions-trading initiatives on which they have been placing big bets."[13]

Public understanding had mostly kept up with the main points of the evolving scientific consensus. Polls in the 1990s found that roughly half of Americans thought global warming was already here, and many of the rest thought it was coming. Fewer than one in eight would assert that it would never happen. But most recognized that they were not well informed. They were easily confused by conflicting statements. When the IPCC's 2001 report concluded that it was likely that greenhouse warming had begun and would get steadily worse, this scarcely seemed like news. What did catch public attention was the new administration's withdrawal from the

Kyoto process. Editorials soundly scolded the action as a surrender to business interests. So it was, but George W. Bush's approach was not far from what a majority of the American public and Congress wanted. To be sure, it would be good to do something about global warming, most people thought—but not if that would mean changing anything very much.

Polls and focus groups in the 1990s found that most people scarcely connected the greenhouse effect with their daily lives. Asked about urgent environmental problems, citizens would bring up immediate concerns such as polluted drinking water, toxic waste, or local smog. Only vaguely understanding the true causes and nature of climate change, the average person could not imagine what practical steps might be taken to forestall it. Some hoped that futuristic technical advances would somehow fix such problems. Others vaguely foresaw a general apocalyptic environmental catastrophe. Almost everyone thought that nothing they personally could do would help. Many were convinced that not only climate changes but all environmental harms were the fault of social decline—a rising tide of selfishness, greed, and corruption. Such people saw a generalized "pollution," the material and moral evils intertwined. Believing they were powerless to halt this deterioration, they considered the problem of global warming insoluble. One study of Americans concluded that most of them, anxious and baffled, literally did not like to think about the subject at all. "Their concern translates into frustration rather than support for action."[14]

While most people only watched with unspoken concern, many scientists, environmental activists, government officials, and even business leaders committed themselves to action. At an international meeting held in Bonn in July 2001, 178 governments—but not the United States—negotiated a compromise agreement for implementing the Kyoto Protocol. Led by Western Europe, nations pledged to enact a variety of measures to restrict greenhouse gas emissions. The goal, watered down from Kyoto, was to return green-house gas emissions to roughly their 1990 rate. Scarcely anyone believed that would really be achieved. And if somehow it did happen, at the 1990 rate of emissions the greenhouse gases in the atmosphere would still continue to rise. The Kyoto Protocol was evidently only a bare beginning for yet more difficult and far-reaching negotiations. Global warming might require the international system to forge entirely new mechanisms of cooperation, and some questioned whether people could rise to the challenge. Many leaders nevertheless felt it worthwhile to keep on developing regulation and monitoring mechanisms. The experience would be essential if the day came when dire need forced the world to truly commit itself to halt global warming.

Far-seeing people in fields ranging from forestry to municipal water supplies began to lay plans for a changed world. More and more experts were confident that they could find practical ways to keep climate change within tolerable limits without harming industrial efficiency. For example, reducing the leakage of methane gas from pipelines would actually save money while significantly reducing one source of global warming. Cutting back on the emission of soot from smokestacks would seriously reduce health problems and their medical expenses, meanwhile reducing the warming that the black particles caused. More generally, less pollution and fewer subsidies for fossil fuels should not weaken the economy but strengthen it even in the short run, as well as for posterity. Meanwhile, people could brace for the changes that were already inevitable.

The chief government activity remained the old standby: another cycle of research. Climate scientists had established that their research deserved substantial funding, and they continued to use international committees to coordinate loosely the way it was spent. At the start of the twenty-first century the world was devoting several billion dollars a year to climate research. That sounded like a lot, yet it was less than was spent on many other scientific and technical problems. It barely sufficed for a subject where the fate of entire populations would be swayed by dozens of factors, each planetary in scope. Long gone were the days when the great questions of climate could be profitably studied by a few people taking time off from their usual research. The job now was to find precise answers to countless specific questions, each requiring teams of highly specialized scientists using costly equipment.

The research was fragmenting because the central question had found its answer. Scientists had reached the end of a grand journey: the discovery of global warming. The hypothesis proposed by Arrhenius in 1896—denied by almost every expert through the first half of the twentieth century and steadily advancing through the second half—was now as well accepted as any scientific proposal of its nature could ever be.

Complete knowledge would come only from watching how the weather actually did evolve. The past two decades of effort had scarcely narrowed the range of uncertainty. Predictions for the doubled CO_2 level expected in half a century or so still stood at roughly 3°C, plus or minus a couple of degrees. Beyond that the geophysics was intractable, with complexities and limitations of data and theory that stubbornly refused to allow more precise conclusions. For a system as complicated and delicate as climate, science could not even exclude the possibility of some shocking surprise, a future sharply different from what the best calculations predicted.

Yet the biggest source of uncertainty now is not in the science. To predict climate change, you would first have to predict changes in CO_2, methane, and other greenhouse gases, plus emissions of smoke and other aerosols, not to mention changes in crops and forests. These changes depend less on geochemistry and biology than on human actions. Whether the world will experience a mild or a drastic warming depends above all on future social and economic trends—population growth, the regulation of soot from smoke-stacks, and so forth. In the third report of the IPCC, scientists have given their best answer. Now the main question is what people will choose to do.

REFLECTIONS

How do scientists get reliable information about the world? When we are told of an advance of science, the language brings up a picture of people marching resolutely ahead. A scientist "discovers" something, like an explorer of old who first comes into an unknown valley. Other explorers push onward, each taking knowledge a step forward. That would be "progress" in the old meaning of the word, a stately parade advancing according to plan.

In reality, after a scientist publishes a paper with an idea or observation, other scientists usually look upon it with justifiable suspicion. Many papers, perhaps most of them, harbor misconceptions or plain errors. After all, research (by definition) operates past the edge of the known. People are peering through fog at a faint shape, never seen before. Every sighting must be checked and confirmed. Scientists find confirmation of an idea all the more convincing when it comes in from the side, using some entirely different type of observation or line of thought. Such connections among different realms are especially common in a science like geophysics, whose subject is intrinsically complex. Scientists may start with something they learned about the smoke from volcanoes, put it alongside telescopic observations of Venus, notice the chemistry of smog in Los Angeles, and plug it all into a computer calculation about clouds. You cannot point to a single observation or model that convinced everyone about anything.

This doesn't look like an exploring team moving into new territory. It looks more like a crowd of people scurrying about, some huddling together to exchange notes, others straining to hear a distant voice or shouting criticism across the hubbub. Everyone is moving in different directions, and it takes a while to see the overall trend. I believe this is the way things commonly proceed, not only in geophysics but in most fields of science.

In this book I have tried to show this process by connecting the dots among roughly a thousand of the most important papers in the science of climate change.[1] For each one of these select thousand, scientists published another ten or so papers of nearly the same importance, describing related data, calculations, or techniques. And for each of those ten thousand, specialists in that particular subject had to scan at least ten other publications that turned out to be less significant—studies that offered minor corroborations, or perhaps contained distracting errors, or that turned out not to be relevant at all. By pulling the main developments above the tumult, this book gives a clearer picture than scientists could see at the time.

Getting coherent explanations is harder in geophysics than in relatively self-contained disciplines like astrophysics or molecular genetics. Scientists in those disciplines address problems that fall within a well-understood boundary. That boundary is roughly congruent with a social boundary, defining a community. The discipline has developed its own journals, scientific societies, meetings, and university departments. Scientists develop these social mechanisms partly to facilitate their work of training students and raising funds for research. Still more, the social coherence of the discipline is invaluable to facilitate their work of communicating findings to one another, debating them, and reaching conclusions about which findings are reliable.

For the process to work, scientists must trust their colleagues. How is the trust maintained? Integrity in telling the truth is important, but it is not enough: while scientists rarely cheat one another, they easily fool themselves. The essential kind of trust comes from sharing a goal, namely, the pursuit of reliable knowledge, and from sharing principles about how to pursue that goal. One necessary principle is to take things apart—tolerating dissent, allowing every rational argument to be heard in public discussion. A second principle is to put things together—arguing out a consensus on important points, even while agreeing to disagree on others.

Maintaining trust is more difficult where the social structure is not cohesive. A community in one specialty cannot thoroughly check the work of researchers in another branch of science, but must accept their word for what is valid. The study of climate change is an extreme example. Researchers cannot isolate meteorology from solar physics, pollution studies from computer science, oceanography from glacier-ice chemistry, and so forth. The range of journals they cite in their footnotes is remarkably broad. This sprawl is inevitable, when so many different factors do in fact influence climate. But the complexity imposes difficulties on those who try to reach solid conclusions abut climate change.

In physics, I can say that a coin will fall at precisely such-and-such an acceleration when you drop it. Not everything can be predicted—the physicist normally cannot tell whether it will land heads or tails—but the general movement is predictable with great exactitude. The reliability of such physical laws can be checked reasonably well by a single person, or at most by one or two teams of physicists. It is otherwise for a question like what the climate will do after we double the amount of CO_2 in the atmosphere. Here we encounter so many nearly chaotic influences that the main facts can be known only roughly. And here establishing the level of reliability depends on a process of checking and correction by dozens of scientific communities, each dealing with its own piece of the problem.

Who made the discovery of global warming—that is to say, the discovery that human activities have begun to make the world warmer? No one person, but a number of scientific communities. Their achievement was not just to accumulate data and perform calculations, but also to link these together. This was obviously a social process, the work of many people interacting with one another. The social process was so complex, and so important, that the last stage was visibly institutionalized: the workshops, reviews, and negotiating sessions of the Intergovernmental Panel on Climate Change. The discovery of global warming was patently a social product, a consensus of judgments arising in countless discussions among thousands of experts.

When people hear of a discovery, they make an implicit assessment of its reliability, that is, how strongly they should believe it is true. The IPCC was pressed to be explicit about the status of its conclusions. When the panel announced in 2001 that they found it "likely" that the current unprecedented rate of warming was largely due to the rise of greenhouse gases, they explained in a footnote what "likely" meant: they judged the probability that the finding was true lay between 66 and 90 percent.[2]

Some skeptics continued to believe that global warming was not likely at all. They pointed insistently at all the places where climate theory was incomplete. And amid the immense volumes of data now available, they found scraps here and there that supported their views. They believed that "global warming" was nothing but a social construction—more like a myth invented by a community than a fact like a rock you could hold in your hand. After all, the critics pointed out, communities of scientists had often held mistaken views, and then changed their collective minds. Hadn't experts once denied that greenhouse warming was possible? Hadn't experts, as recently as the 1970s, warned of a new ice age?

Most scientists found this denial not persuasive—worse, hardly even interesting. The critics' data and arguments looked feeble when they were set against the enormous mass of evidence for greenhouse warming. To be sure, half a century back, most scientists had found Callendar's greenhouse-warming proposition implausible. But scientists back then had understood that their ideas about climate change were based on no more than a scattering of uncertain measurements and hand-waving arguments. Callendar's proposition, although it flew in the face of ideas about climate stability that scientists had long taken for granted, was only provisionally set aside. It stuck in the minds of the experts, awaiting the coming of better data and theories. Likewise, during the controversies of the 1970s, most scientists had explained time and again that their knowledge was still too primitive to say whether the climate would turn warm or cold. Their main point was that they had now learned enough to give up the old confidence in stability. By the end of the twentieth century it was the critics, arguing for a self-regulating climate, who were struggling to maintain the traditional belief.

By that time not only scientists but most people had reluctantly reached a less comfortable view of the natural world and its relationship with human civilization. The views of the public and of the scientific community had necessarily changed together, each acting upon the other. On the public side, hard experience drove home how severely our technologies can change everything, even the air itself. Meanwhile, on the scientific side, knowledge of how climate could change evolved under the influence of countless field observations, laboratory measurements, and numerical calculations, yet within limits set by the larger community's common-sense understanding (and by the funding it provided). Eventually, the conclusions were solid enough to be perpetuated in consensus panel reports. Certainly, in a restricted sense, one could call the resulting understanding of climate change a product of human society.

We should not call it *nothing but* a social product. Future climate change in this regard is like electrons, galaxies, and many other things not immediately accessible to our senses. All these concepts emerged from a vigorous struggle of ideas, until most people were persuaded to say the concepts represented something real. Just what "real" meant was open to debate; philosophers offered many opinions about how a scientific concept might somehow correspond to an ultimate reality. That ageless question rarely troubled climate scientists, who took it for granted that the future climate is as real as a rock. At the same time, scientists readily admitted that their knowledge of this future thing could only be stated within a range of probabilities.

Our understanding of climate goes beyond scientific reports into a wider realm of thinking. When I look at a snowless street in January I may see a natural weather variation, or I may see a human artifact caused by greenhouse gas emissions. Such perceptions are shaped not only by scientists, but by interest groups, politicians, and the media. For global warming in particular, the social influences run deeper still. Unlike, say, the orbits of planets, the climate in the future actually does depend in part on what we think about it. For what we think will determine what we do.

Faced with scientists who publish warnings, the public's natural response is to ask them for definitive guidance. When the scientists fail to say for certain what will happen, politicians habitually tell them to go back and do more research. That is all very well, but in the case of climate, waiting for a sure answer would mean waiting forever. When we are faced with a new disease or an armed invasion, we do not put off decisions until more research is done: we act using the best guidelines available.

✳ What can we do about global warming, and what should we do?

My training as a physicist and historian of science has given me some feeling for where scientific claims are reliable and where they are shaky. Of course climate science is full of uncertainties, and nobody claims to know exactly what the climate will do. That very uncertainty is part of what, I am confident, is known beyond doubt: our planet's climate can change, tremendously and unpredictably. Beyond that we can conclude (with the IPCC) that it is *very likely* that significant global warming is coming in our lifetimes. This surely brings a likelihood of harm, widespread and grave. The few who contest these facts are either ignorant or so committed to their viewpoint that they will seize on any excuse to deny the danger.

Thanks to the strenuous labors by thousands of people described in these pages, we have had a warning in time—although just barely in time. If there is even a small risk that your house will burn down, you will take care to install smoke alarms and buy insurance. We can scarcely do less for the well-being of our society and the planet's ecosystems. Thus the only useful discussion is over what measures are worth their cost.

Many things can be done right now that are not only cheap and effective, but will actually pay for themselves through benefits entirely aside from acting against global warming. Americans in particular—the world's most promiscuous emitters of greenhouse gases and the ones best placed to do something about it—must set an example. A good start would be to remove the government subsidies for fossil fuels, which are huge, mostly hidden, and economically unsound. Another sensible step would be to raise the tax on gasoline by a few dollars (comparable to what nearly all other industrial nations pay, and compensated by lowering other taxes) to cover the actual costs of roads, traffic congestion, and medical care for accident injuries and illness due to smog. Other economically beneficial policies could improve fuel efficiency in many areas, protect forests, and so forth. Looking beyond CO_2, money can actually be saved while reducing the greenhouse effect by fixing leakage of methane from pipelines, attacking unhealthy smoke emissions, and carrying out similar changes. Such steps can be taken not only by national governments but by local governments, and by most businesses and individual citizens.

Most important of all, regulation and "price signals" will stimulate development of technologies and practices that can advance human welfare with far lower greenhouse gas emission. A good bit of that development is already under way, but technologies do not magically grow by themselves. According to economic demands, technology may remain stagnant or dash forward to solve problems with remarkable speed. The control of CFCs, for example, has turned out to be far easier and cheaper than the regulated industries feared.

To say that such steps are socially or politically impossible is to forget that far greater changes have come swiftly, in countless areas, once people set their minds to it (think how Americans' patterns of living, even of eating, have changed over the past 50 years!). Citizens can reconsider their personal practices and put pressure on businesses and governments. This is not a job for someone else, sometime down the road: we have already run out of time. Without delay, nations should join—as nearly all but the United States have done—in working out systems for applying standards on the international scale, which is where climate operates. The first practical steps, the really cheap and easy ones, will not have a big impact on future global warming. But starting off will give the world experience in developing and negotiating the right technologies and policies. We will need this experience if, as is likely, climate change becomes so harmful that we are compelled into much greater efforts.

Like many threats, global warming calls for greater government activity, and that rightly worries people. But in the twenty-first century the alternative to government action is not individual liberty; it is corporate power. And the role of large corporations in this story has been mostly negative, a tale of self-interested obfuscation and short-sighted delay. The atmosphere is a classic case of a "commons": in the old shared English meadow, any given individual was bound to gain by adding more of his own cows, although everyone lost from the overgrazing. In such cases the public interest can only be protected by public rules.

Much more likely than not, global warming is upon us. We should expect weather patterns to continue to change and the seas to continue to rise, in an ever worsening pattern, in our lifetimes and on into our grandchildren's. The question has graduated from the scientific community: climate change is a major social, economic, and political issue. Nearly everyone in the world will need to adjust. It will be hardest for the poorer groups and nations among us, but nobody is exempt. Citizens will need reliable information, the flexibility to change their personal lives, and efficient and appropriate help from all levels of government. So it is an important job, in some ways our top priority, to improve the communication of knowledge and to strengthen democratic control in governance everywhere. The spirit of fact-gathering, rational discussion, toleration of dissent, and negotiation of an evolving consensus, which has characterized the climate science community, can serve well as a model.

1800–1870   Level of carbon dioxide gas (CO_2) in the atmosphere, as later measured in ancient ice, is about 290 ppm (parts per million).

First Industrial Revolution. Coal, railroads, and land clearing speed up greenhouse gas emission, while better agriculture and sanitation speed up population growth.

1896   Arrhenius publishes first calculation of global warming from human emissions of CO_2.

1897   Chamberlin produces a model for global carbon exchange including feedbacks.

1870–1910   Second Industrial Revolution. Fertilizers and other chemicals, electricity, and public health further accelerate growth.

1914–1918   World War I; governments learn to mobilize and control industrial societies.

1920–1925   Opening of Texas and Persian Gulf oil fields inaugurates era of cheap energy.

1930s   Global warming trend since late 19th century reported.

Milankovitch proposes orbital changes as the cause of Ice Ages.

1938   Callendar argues that CO_2 greenhouse global warming is under way, reviving interest in the question.

1939–1945   World War II. Grand strategy is largely driven by a struggle to control oil fields.

1945   U.S. Office of Naval Research begins generous funding of many fields of science, some of which happen to be useful for understanding climate change.

1956   Ewing and Donn offer a feedback model for quick ice-age onset.

Phillips produces a somewhat realistic computer model of the global atmosphere.

Plass calculates that adding CO_2 to the atmosphere will have a significant effect on the radiation balance.

1957   Launch of Soviet Sputnik satellite. Cold War concerns support 1957–58 International Geophysical Year, bringing new funding and coordination to climate studies.

Revelle finds that CO_2 produced by humans will not be readily absorbed by the oceans.

1958   Telescope studies show a greenhouse effect raises temperature of the atmosphere of Venus far above the boiling point of water.

1960   Downturn of global temperatures since the early 1940s is reported.

Keeling accurately measures CO_2 in the Earth's atmosphere and detects an annual rise. The level is 315 ppm.

1962   Cuban Missile Crisis, peak of the Cold War.

1963   Calculations suggest that feedback with water vapor could make the climate acutely sensitive to changes in CO_2 level.

1965   Boulder meeting on causes of climate change, in which Lorenz and others point out the chaotic nature of the climate system and the possibility of sudden shifts.

1966   Emiliani's analysis of deep-sea cores shows the timing of ice ages was set by small orbital shifts, suggesting that the climate system is sensitive to small changes.

1967   International Global Atmospheric Research Program established, mainly to gather data for better short-range weather prediction but including climate.

Manabe and Wetherald make a convincing calculation that doubling CO_2 would raise world temperatures a couple of degrees.

1968   Studies suggest a possibility of collapse of Antarctic ice sheets, which would raise sea levels catastrophically.

1969   Astronauts walk on the Moon, and people perceive the Earth as a fragile whole.

Budyko and Sellers present models of catastrophic ice-albedo feedbacks.

Nimbus III satellite begins to provide comprehensive global atmospheric temperature measurements.

1970   First Earth Day. Environmental movement attains strong influence, spreads concern about global degradation.

Creation of U.S. National Oceanic and Atmospheric Administration, the world's leading funder of climate research.

Aerosols from human activity are shown to be increasing swiftly. Bryson claims they are causing global cooling.

1971   SMIC conference of leading scientists reports a danger of rapid and serious global climate change caused by humans, calls for an organized research effort.

Mariner 9 spacecraft finds a great dust storm warmed the atmosphere of Mars, plus indications of a radically different climate in the past.

1972   Ice cores and other evidence show big climate shifts in the past between relatively stable modes in the space of a thousand years or so.

1973   Oil embargo and price rise bring first "energy crisis."

1974   Serious droughts and other unusual weather since 1972, plus warnings by scientists and journalists, raise public concern about climate change, perhaps a new ice age.

1975   Concern about environmental effects of airplanes leads to investigations of trace gases in the stratosphere and discovery of danger to ozone layer.

Manabe and collaborators produce complex but plausible computer models which show a temperature rise of several degrees for doubled CO_2.

1976   Studies find that CFCs (1975) and methane and ozone (1976) can make a serious contribution to the greenhouse effect.

Deep-sea cores show a dominating influence from 100,000-year Milankovitch orbital changes, emphasizing the role of feedbacks.

Deforestation and other ecosystem changes are recognized as major factors in the future of the climate.

Eddy shows that there were prolonged periods without sunspots in past centuries, corresponding to cold periods.

1977   Scientific opinion tends to converge on rapid global warming as the biggest climate risk.

1978   Attempts to coordinate climate research in U.S. end with an inadequate National Climate Program Act, accompanied by rapid but temporary growth in funding.

1979   Second oil "energy crisis." Strengthened environmental movement encourages renewable energy sources, inhibits nuclear energy growth.

U.S. National Academy of Sciences report finds it highly credible that doubling CO_2 will bring 1.5–4.5°C global warming.

World Climate Research Program launched to coordinate international research.

Election of President Reagan brings backlash against environmental movement; political conservatism is linked to skepticism about global warming.

1981   IBM Personal Computer introduced. Advanced economies are increasingly delinked from energy.

Hansen and others show that sulfate aerosols can significantly cool the climate, raising confidence in models showing future greenhouse warming.

Some scientists predict greenhouse warming "signal" should be visible by about the year 2000.

1982   Greenland ice cores reveal drastic temperature oscillations in the space of a century in the distant past.

Strong global warming since mid-1970s is reported, with 1981 the warmest year on record.

1983   Reports from U.S. National Academy of Sciences and Environmental Protection Agency spark conflict; greenhouse warming becomes prominent in mainstream politics.

1985   Villach Conference declares consensus among experts that some global warming seems inevitable and international agreements to restrict emissions should be considered.

Antarctic ice cores show that CO_2 and temperature went up and down together through past ice ages.

Broecker speculates that a reorganization of North Atlantic Ocean circulation can bring swift and radical climate change.

1987   Montreal Protocol of the Vienna Convention imposes international restrictions on emission of ozone-destroying gases.

1988   News media coverage of global warming leaps upward following record heat and droughts plus testimony by Hansen.

Toronto Conference calls for strict, specific limits on greenhouse gas emissions.

Ice-core and biology studies confirm living ecosystems make climate feedback by way of methane, which could accelerate global warming.

Intergovernmental Panel on Climate Change (IPCC) is established.

Level of CO_2 in the atmosphere reaches 350 ppm.

The period since 1988 is too recent to identify historical milestones.

For complete references see http://www.aip.org/history/climate

1. How Could Climate Change?

1. "Warmer World," *Time*, 2 Jan. 1939, p. 27.
2. Albert Abarbanel and Thomas McCluskey, "Is the World Getting Warmer?" *Saturday Evening Post*, 1 July 1950, p. 63.
3. "Warmer World," p. 27.
4. G. S. Callendar, "The Artificial Production of Carbon Dioxide and Its Influence on Climate," *Quarterly J. Royal Meteorological Society* 64 (1938): 223–240.
5. John Tyndall, "Further Researches on the Absorption and Radiation of Heat by Gaseous Matter" (1862), in Tyndall, *Contributions to Molecular Physics in the Domain of Radiant Heat* (New York: Appleton, 1873), p. 117.
6. John Tyndall, "On Radiation through the Earth's Atmosphere," *Philosophical Magazine* ser. 4, 25 (1863): 204–205.
7. Athelstan Spilhaus, interview by Ron Doel, November 1989, American Institute of Physics, College Park, Md.
8. H. Lamb quoted in Tom Alexander, "Ominous Changes in the World's Weather," *Fortune*, Feb. 1974, p. 90.
9. William Joseph Baxter, *Today's Revolution in Weather* (New York: International Economic Research Bureau, 1953), p. 69.
10. Thomas C. Chamberlin, "On a Possible Reversal of Deep-Sea Circulation and Its Influence on Geologic Climates," *J. Geology* 14 (1906): 371.
11. James R. Fleming, *Historical Perspectives on Climate Change* (New York: Oxford University Press, 1998), chaps. 2–4.
12. Hubert H. Lamb, *Through All the Changing Scenes of Life: A Meteorologist's Tale* (Norfolk, UK: Taverner, 1997), pp. 192–193.

2. Discovering a Possibility

1. C.-G. Rossby, "Current Problems in Meteorology," in *The Atmosphere and the Sea in Motion*, ed. Bert Bolin (New York: Rockefeller Institute Press, 1959), p. 15.

2. Interview of Plass by Weart, 14 March 1996, American Institute of Physics, College Park, Md.

3. G. N. Plass, "Carbon Dioxide and the Climate," *American Scientist* 44 (1956): 302–316.

4. Roger Revelle, "The Oceans and the Earth," talk at American Association for the Advancement of Sciences symposium, Dec. 27, 1955, typescript, folder 66, Box 28, Revelle Papers MC6, Scripps Institution of Oceanography archives, La Jolla, Calif.

5. Roger Revelle and Hans E. Suess, "Carbon Dioxide Exchange between Atmosphere and Ocean and the Question of an Increase of Atmospheric CO_2 During the Past Decades," *Tellus* 9 (1957): 18–27.

6. Clark A. Miller, "Scientific Internationalism in American Foreign Policy: The Case of Meteorology, 1947–1958," in *Changing the Atmosphere. Expert Knowledge and Environmental Governance,* ed. Clark A. Miller and Paul N. Edwards (Cambridge, Mass.: MIT Press, 2001), p. 171 and *passim.*

7. J. A. Eddy, interview by Weart, April 1999, American Institute of Physics, College Park, Md., p. 4.

8. C. C. Wallén, "Aims and Methods in Studies of Climatic Fluctuations," in *Changes of Climate. Proceedings of the Rome Symposium Organized by UNESCO and the World Meteorological Organization, 1961* (UNESCO Arid Zone Research Series, 20) (Paris: UNESCO, 1963), p. 467.

9. Roger Revelle, interview by Earl Droessler, Feb. 1989, American Institute of Physics, College Park, Md.

10. Charles D. Keeling, "The Concentration and Isotopic Abundances of Carbon Dioxide in the Atmosphere," *Tellus* 12 (1960): 200–203.

3. A Delicate System

1. Jhan and June Robbins, "100 Years of Warmer Weather," *Science Digest,* Feb. 1956, p. 83.

2. Helmut Landsberg, reported in the *New York Times,* Feb. 15, 1959.

3. United States Congress (85:2), House of Representatives, Committee on Appropriations, *Report on the International Geophysical Year* (Washington, D.C.: Government Printing Office, 1957), pp. 104–106.

4. J. Gordon Cook, *Our Astonishing Atmosphere* (New York: Dial, 1957), p. 121.

5. The Conservation Foundation, *Implications of Rising Carbon Dioxide Content of the Atmosphere* (New York: The Conservation Foundation, 1963).

6. National Academy of Sciences, Committee on Atmospheric Sciences Panel on Weather and Climate Modification, *Weather and Climate Modification: Problems and Prospects.* 2 vols. (Washington, D.C.: National Academy of Sciences, 1966), vol. 1, p. 10.

7. President's Science Advisory Committee, *Restoring the Quality of Our Environment. Report of the Environmental Pollution Panel* (Washington, D.C.: The White House, 1965), p. 26.

8. National Academy of Sciences, *Weather and Climate Modification,* vol. 1, pp. 16, 20.

9. Cesare Emiliani, "Ancient Temperatures," *Scientific American,* Feb. 1958, p. 54.

10. Kenneth J. Hsü, *Challenger at Sea: A Ship That Revolutionized Earth Science* (Princeton, N.J.: Princeton University Press, 1992), pp. 30–32, 220.

11. Wallace S. Broecker, "In Defense of the Astronomical Theory of Glaciation," *Meteorological Monographs* 8, no. 30 (1968): 139.

12. Broecker et al., "Milankovitch Hypothesis Supported by Precise Dating of Coral Reef and Deep-Sea Sediments," *Science* 159 (1968): 300.

13. C. E. P. Brooks, "The Problem of Mild Polar Climates," *Quarterly J. Royal Meteorological Society* 51 (1925): 90–91.

14. David B. Ericson et al., "Late-Pleistocene Climates and Deep-Sea Sediments," *Science* 124 (1956): p. 388.

15. Broecker, "Application of Radiocarbon to Oceanography and Climate Chronology." PhD Thesis, Columbia University, 1957, pp. V–9.

16. Harry Wexler, "Variations in Insolation, General Circulation and Climate," *Tellus* 8 (1956): 480.

17. Broecker, interview by Weart, Nov. 1997, American Institute of Physics, College Park, Md.

18. Lewis F. Richardson, *Weather Prediction by Numerical Process* (Cambridge: Cambridge University Press, 1922; rpt. New York: Dover, 1965), pp. 219, ix.

19. Jule G. Charney et al., "Numerical Integration of the Barotropic Vorticity Equation," *Tellus* 2 (1950): 245.

20. C.-G. Rossby, "Current Problems in Meteorology," in *The Atmosphere and the Sea in Motion,* ed. Bert Bolin (New York: Rockefeller Institute Press, 1959), p. 30.

21. Norbert Wiener, "Nonlinear Prediction and Dynamics," in *Proceedings of the Third Berkeley Symposium on Mathematical Statistics and Probability,* ed. Jerzey Neyman (Berkeley: University of California Press, 1956), p. 247.

22. Edward N. Lorenz, "Deterministic Nonperiodic Flow," *J. Atmospheric Sciences* 20 (1963): 130, 141.

23. Lorenz, "Climatic Determinism," *Meteorological Monographs* 8 (1968): 3.

24. J. Murray Mitchell, "Concluding Remarks" [based on Revelle's summary at the conference], in Mitchell, "Causes of Climatic Change" (*Proceedings,* VII Congress, International Union for Quaternary Research, vol. 5, 1965), *Meteorological Monographs* 8, no. 30 (1968): 157–158.

25. Hubert H. Lamb, "Climatic Fluctuations," in *General Climatology,* ed. H. Flohn (Amsterdam: Elsevier, 1969), p. 178.

4. A Visible Threat

1. Reid A. Bryson and Wayne M. Wendland, "Climatic Effects of Atmospheric Pollution," in *Global Effects of Environmental Pollution,* ed. S. F. Singer (New York: Springer-Verlag, 1970), p. 137.

2. J. Murray Mitchell, Jr., "Recent Secular Changes of Global Temperature," *Annals of the New York Academy of Sciences* 95 (1961): 247.

3. SCEP (Study of Critical Environmental Problems), *Man's Impact on the Global Environment. Assessment and Recommendation for Action* (Cambridge, Mass.: MIT Press, 1970), p. 12.

4. Carroll L. Wilson and William H. Matthews, eds., *Inadvertent Climate Modification. Report of Conference, Study of Man's Impact on Climate (SMIC), Stockholm* (Cambridge, Mass.: MIT Press, 1971), pp. 129, v.

5. David A. Barreis and Reid A. Bryson, "Climatic Episodes and the Dating of the Mississippian Cultures," *Wisconsin Archeologist,* Dec. 1965, p. 204.

6. Bryson, "A Perspective on Climatic Change," *Science* 184 (1974): 753–760; Bryson et al., "The Character of Late-Glacial and Postglacial Climatic Changes (Symposium, 1968)," in *Pleistocene and Recent Environments of the Central Great Plains (University of Kansas Department of Geology, Special Publication),* ed. Wakefield Dort, Jr. and J. Knox Jones, Jr. (Lawrence: University of Kansas Press, 1970), p. 72; W. M. Wendland and Bryson, "Dating Climatic Episodes of the Holocene," *Quaternary Research* 4 (1974): 9–24.

7. Richard B. Alley, *The Two-Mile Time Machine* (Princeton, N.J.: Princeton University Press, 2000); Paul A. Mayewski and Frank White, *The Ice Chronicles: The Quest to Understand Global Climate Change* (Hanover, N.H.: University Press of New England, 2002).

8. Mitchell, "The Natural Breakdown of the Present Interglacial and Its Possi-

ble Intervention by Human Activities," *Quaternary Research* 2 (1972): 437–438.

9. W. Dansgaard et al., "Speculations about the Next Glaciation," *Quaternary Research* 2 (1972): 396.

10. Johannes Weertman, "Stability of the Junction of an Ice Sheet and an Ice Shelf," *J. Glaciology* 13 (1974): 3.

11. George J. Kukla and R. K. Matthews, "When Will the Present Interglacial End?" *Science* 178 (1972): 190–191.

12. Christian E. Junge, "Atmospheric Chemistry," *Advances in Geophysics* 5 (1958): 95.

13. Mitchell, "A Preliminary Evaluation of Atmospheric Pollution as a Cause of the Global Temperature Fluctuation of the Past Century," in *Global Effects of Environmental Pollution,* ed. S. Fred Singer (New York: Springer-Verlag, 1970), p. 153.

14. S. Ichtiaque Rasool and Stephen H. Schneider, "Atmospheric Carbon Dioxide and Aerosols: Effects of Large Increases on Global Climate," *Science* 173 (1971): 138.

15. G. D. Robinson, "Review of Climate Models," in *Man's Impact on the Climate* [Study of Critical Environmental Problems (SCEP) Report], ed. William H. Matthews et al.(Cambridge, Mass.: MIT Press, 1971), p. 214.

16. Mikhail I. Budyko, "The Effect of Solar Radiation Variations on the Climate of the Earth," *Tellus* 21 (1969): 618.

17. William D. Sellers, "A Global Climatic Model Based on the Energy Balance of the Earth-Atmosphere System," *J. Applied Meteorology* 8 (1969): 392.

18. Andrew P. Ingersoll, "The Runaway Greenhouse: A History of Water on Venus," *J. Atmospheric Sciences* 26 (1969): 1191–1198; S. Ichtiaque Rasool and Catheryn de Bergh, "The Runaway Greenhouse and the Accumulation of CO_2 in the Venus Atmosphere," *Nature* 226 (1970): 1037–1039.

19. Owen B. Toon et al., "Climatic Change on Mars and Earth," in *Proceedings of the WMO/IAMAP Symposium on Long-Term Climatic Fluctuations, Norwich, Aug. 1975* (WMO Doc. 421) (Geneva: World Meteorological Organization, 1975), p. 495.

5. Public Warnings

1. William A. Reiners, "Terrestrial Detritus and the Carbon Cycle," in *Carbon and the Biosphere,* ed. George M. Woodwell and Erene V. Pecan (Washing-

ton, D.C.: Atomic Energy Commission [National Technical Information Service, CONF-7502510], 1973), p. 327.

2. Tom Alexander, "Ominous Changes in the World's Weather," *Fortune,* Feb. 1974, p. 92.

3. "Another Ice Age?" *Time,* 26 June 1974, p. 86.

4. G. S. Benton quoted in *New York Times,* April 30, 1970.

5. Lowell Ponte, *The Cooling* (Englewood Cliffs, N.J.: Prentice-Hall, 1976), pp. 234–235.

6. "The Weather Machine," BBC-television and WNET, expanded in a book: Nigel Calder, *The Weather Machine* (New York: Viking, 1975), quote p. 134.

7. John Gribbin, "Man's Influence Not Yet Felt by Climate," *Nature* 264 (1976): 608; B. J. Mason, "Has the Weather Gone Mad?" *The New Republic,* 30 July 1977, pp. 21–23.

8. Reid A. Bryson and Thomas J. Murray, *Climates of Hunger: Mankind and the World's Changing Weather* (Madison: University of Wisconsin Press, 1977).

9. Stephen H. Schneider with Lynne E. Mesirow, *The Genesis Strategy: Climate and Global Survival* (New York: Plenum Press, 1976), esp. chap. 3.

10. Gerald Stanhill, "Climate Change Science Is Now Big Science," *Eos, Transactions of the American Geophysical Union* 80, no. 35 (1999): 396 (from graph).

11. National Academy of Sciences, Committee on Atmospheric Sciences, Panel on Weather and Climate Modification, *Weather and Climate Modification: Problems and Prospects.* 2 vols. (Washington, D.C.: National Academy of Sciences, 1966), vol. 1, p. 11.

12. E.g., E. P. Stebbing, "The Encroaching Sahara: The Threat to the West African Colonies," *Geographical J.* 85 (1935): 523.

13. Charles D. Keeling, "The Carbon Dioxide Cycle: Reservoir Models to Depict the Exchange of Atmospheric Carbon Dioxide with the Ocean and Land Plants," in *Chemistry of the Lower Atmosphere,* ed. S. I. Rasool (New York: Plenum, 1973), p. 320.

14. Ibid., p. 279.

15. George M. Woodwell, "The Carbon Dioxide Question," *Scientific American,* Jan. 1978, p. 43.

16. Wallace S. Broecker et al., "Fate of Fossil Fuel Carbon Dioxide and the Global Carbon Budget," *Science* 206 (1979): 409, 417.

17. Joseph Smagorinsky, "Numerical Simulation of the Global Circulation," in

Global Circulation of the Atmosphere, ed. G. A. Corby (London: Royal Meteorological Society, 1970), p. 33.

18. Broecker, interview by Weart, Nov. 1997, American Institute of Physics, College Park, Md.

19. P. H. Abelson, "Energy and Climate," *Science* 197 (1977): 941.

20. "CO_2 Pollution May Change the Fuel Mix," *Business Week,* 8 Aug. 1977, p. 25; "The World's Climate Is Getting Worse," *Business Week,* 2 Aug. 1976, p. 49.

21. The "Charney report," National Academy of Sciences, Climate Research Board, *Carbon Dioxide and Climate: A Scientific Assessment* (Washington, D.C.: National Academy of Sciences, 1979), pp. 2, 3; Nicholas Wade, "CO_2 in Climate: Gloomsday Predictions Have No Fault," *Science* 206 (1979): 912–913.

22. Opinion Research Corporation polls, May 1981, USORC.81MAY.R22 and April 1980, USORC.80APR1.R3M. Data furnished by Roper Center for Public Opinion Research, Storrs, Conn.

6. The Erratic Beast

1. Address by Lorenz to the American Association for the Advancement of Science, Washington, D.C., Dec. 29, 1979.

2. James E. Hansen et al., "Climate Impact of Increasing Atmospheric Carbon Dioxide," *Science* 213 (1981): 961.

3. National Academy of Sciences, Climate Research Board, *Carbon Dioxide and Climate: A Scientific Assessment* (Washington, D.C.: National Academy of Sciences, 1979), p. 2.

4. Hansen, "Climate Impact," p. 957; Roland A. Madden and V. Ramanathan, "Detecting Climate Change Due to Increasing Carbon Dioxide," *Science* 209 (1980): 763–768.

5. Stephen H. Schneider, "Introduction to Climate Modeling," in *Climate System Modeling,* ed. Kevin E. Trenberth (Cambridge: Cambridge University Press, 1992), p. 26.

6. Wallace S. Broecker, "Climatic Change: Are We on the Brink of a Pronounced Global Warming?" *Science* 189 (1975): 460–464.

7. Hans E. Suess, "Climatic Changes, Solar Activity, and the Cosmic-Ray Production Rate of Natural Radiocarbon," *Meteorological Monographs* 8, no. 30 (1968): 146.

8. R. E. Dickinson, "Solar Variability and the Lower Atmosphere," *Bulletin of the American Meteorological Society* 56 (1975): 1240–1248.

9. Eddy, interview by Weart, April 1999, American Institute of Physics, College Park, Md.

10. Jack A. Eddy, "Historical Evidence for the Existence of the Solar Cycle," in *The Solar Output and Its Variation,* ed. Oran R. White (Boulder, Colo.: Colorado Associated University Press, 1977), p. 69.

11. Raymond S. Bradley, *Quaternary Paleoclimatology: Methods of Paleoclimatic Reconstruction* (Boston: Allen & Unwin, 1985), p. 69.

12. National Academy of Sciences, United States Committee for the Global Atmospheric Research Program (GARP), *Understanding Climatic Change: A Program for Action* (Washington, D.C.: National Academy of Sciences, 1975), p. 4.

13. Kirk Bryan, "Climate and the Ocean Circulation. III. The Ocean Model," *Monthly Weather Review* 97 (1969): 822.

14. R. O. Reid et al., *Numerical Models of World Ocean Circulation* (Washington, D.C.: National Academy of Sciences, 1975), p. 3.

15. James E. Hansen et al., "Climate Response Times: Dependence on Climate Sensitivity and Ocean Mixing," *Science* 229 (1985): 857–859.

16. W. Dansgaard et al., "A New Greenland Deep Ice Core," *Science* 218 (1982): 1273.

17. U. Siegenthaler et al., "Lake Sediments as Continental Delta O^{18} Records from the Glacial/Post-Glacial Transition," *Annals of Glaciology* 5 (1984): 149.

18. Broecker, "The Biggest Chill," *Natural History,* Oct. 1987, 74–82, p. 87; Broecker et al., "Does the Ocean-Atmosphere System Have More Than One Stable Mode of Operation?" *Nature* 315 (1985): 21–25.

19. Broecker, "The Biggest Chill," p. 82.

7. Breaking into Politics

1. Albert Gore, Jr., *Earth in the Balance: Ecology and the Human Spirit* (Boston: Houghton Mifflin, 1992), pp. 4–6.

2. Robert G. Fleagle, "The U.S. Government Response to Global Change: Analysis and Appraisal," *Climatic Change* 20 (1992): 72.

3. James E. Jensen, "An Unholy Trinity: Science, Politics and the Press" (unpublished talk), 1990.

4. Aug. 22, 1981, p. 1, and Aug. 29, 1981, p. 22.

5. National Academy of Sciences, Carbon Dioxide Assessment Committee, *Changing Climate* (Washington, D.C.: National Academy of Sciences, 1983), p. 3.

6. Stephen Seidel and Dale Keyes, *Can We Delay a Greenhouse Warming?* (Washington, D.C.: Environmental Protection Agency, 2nd ed., 1983), pp. ix, 7 (of sect. 7).

7. Bert Bolin et al., eds., *The Greenhouse Effect, Climatic Change, and Ecosystems. SCOPE Report No. 29.* (Chichester: John Wiley, 1986), pp. xx–xxi.

8. Jonathan Weiner, *The Next One Hundred Years: Shaping the Fate of Our Living Earth* (New York: Bantam, 1990), p. 79.

9. Stephen H. Schneider, "An International Program on 'Global Change': Can It Endure?" *Climatic Change* 10 (1987): 215.

10. *New York Times,* 24 June 1988, p. 1.

11. Spencer R. Weart, *Never at War: Why Democracies Will Not Fight One Another* (New Haven: Yale University Press, 1998), p. 265.

8. The Discovery Confirmed

1. Published surveys include Stanley A. Chagnon et al., "Shifts in Perception of Climate Change: A Delphi Experiment Revisited," *Bulletin of the American Meteorological Society* 73, no. 10 (1992): 1623–1627, and David H. Slade, "A Survey of Informal Opinion Regarding the Nature and Reality of a 'Global Greenhouse Warming,'" *Climatic Change* 16 (1990): 1–4.

2. Frederick Seitz, ed., *Global Warming Update: Recent Scientific Findings* (Washington, D.C.: George C. Marshall Institute, 1992), p. 28.

3. L. Roberts, "Global Warming: Blaming the Sun," *Science* 246 (1989): 992–993.

4. Robert Lichter, "A Study of National Media Coverage of Global Climate Change 1985–1991" (Washington, D.C.: Center for Science, Technology & Media, 1992).

5. *New York Times,* April 19, 1990, p. B4.

6. Tom M. L. Wigley, "Outlook Becoming Hazier," *Nature* 369 (1994): 709–710.

7. Intergovernmental Panel on Climate Change, *Climate Change 1995: The Science of Climate Change,* ed. J. T. Houghton et al. (Cambridge: Cambridge University Press, 1996), online at http://www.ipcc.ch/pub/reports.htm; Richard A. Kerr, "It's Official: First Glimmer of Greenhouse Warming Seen," *Science* 270 (1995): 1565–1567. The press quoted a prelim-

inary version. The final consensus phrasing was, "the observed warming trend is unlikely to be completely natural in origin," Intergovernmental Panel, *Climate Change 1995,* p. 22.

8. Wigley and P. M. Kelly, "Holocene Climatic Change, ^{14}C Wiggles and Variations in Solar Irradiance," *Philosophical Transactions of the Royal Society of London* A330 (1990): 558.

9. Wallace S. Broecker, "Thermohaline Circulation, the Achilles Heel of Our Climate System: Will Man-Made CO_2 Upset the Current Balance?" *Science* 278 (1997): 1582-1588.

10. Intergovernmental Panel, *Climate Change 1995,* p. 7.

11. Intergovernmental Panel, *Climate Change 2001: The Scientific Basis. Contribution of Working Group I to the Third Assessment Report of the IPCC,* ed. J. T. Houghton et al. (Cambridge: Cambridge University Press, 2001), online at http://www.ipcc.ch/pub/reports.htm.

12. Broecker, "Thermohaline Circulation," p. 1586.

13. "Beyond the Hague," *Economist,* 2 Dec. 2000, p. 20; see also p. 61.

14. John Immerwahr, *Waiting for a Signal: Public Attitudes toward Global Warming, the Environment and Geophysical Research* (New York: Public Agenda, 1999), online at http://Earth.agu.org/sci_soc/sci_soc.html; summary in Randy Showstock, "Report Suggests Some Public Attitudes about Geophysical and Environmental Issues," *Eos, Transactions of the American Geophysical Union* 80, no. 24 (1999): 269, 276.

Reflections

1. The bibliography is at http://www.aip.org/history/climate/bib.htm.

2. Intergovernmental Panel on Climate Change, *Climate Change 2001: The Scientific Basis. Contribution of Working Group I to the Third Assessment Report of the IPCC,* ed. J. T. Houghton et al. (Cambridge: Cambridge University Press, 2001), pp. 1, 6, 8, 13, online at http://www.ipcc.ch/pub/reports.htm.

To explore the history further, I naturally recommend first my Web site, which holds about three times as much material as this book: http://www.aip.org/history/climate. The books and articles listed below are also recommended, although with reservations. Not much has been written on the history of climate science. Little of that was written by professional historians, and they mostly stuck to one or another of the many specialized topics. Reviews by scientists are technical, although nearly always accurate; writings by journalists are readable but not everywhere reliable. The full bibliography for this book, in particular the original scientific papers, will be found on my Web site.

Christianson, Gale E. 1999. Greenhouse: The 200-year story of global warming. New York: Walker.

Edwards, Paul N. 2000. "A brief history of atmospheric general circulation modeling." In General circulation model development, edited by D. A. Randall. San Diego, CA: Academic Press.

Fleagle, Robert G. 1992. "From the International Geophysical Year to global change." Reviews of Geophysics 30:305–13.

Fleming, James R. 1998. Historical perspectives on climate change. New York: Oxford University Press.

Handel, Mark David, and James S. Risbey. 1992. "An annotated bibliography on the greenhouse effect and climate change." Climatic Change 21:97–255.

Houghton, John. 1997. Global warming: The complete briefing. Cambridge: Cambridge University Press. 2nd ed.

Imbrie, John, and Katherine Palmer Imbrie. 1986. Ice ages: Solving the mystery. Rev. ed. Cambridge, MA: Harvard University Press.

Kellogg, William W. 1987. "Mankind's impact on climate: The evolution of an awareness." Climatic Change 10:113–36.

Miller, Clark A., and Paul N. Edwards, eds. 2001. *Changing the atmosphere. Expert knowledge and environmental governance.* Cambridge, MA: MIT Press.

Nebeker, Frederik. 1995. *Calculating the weather: Meteorology in the 20th century.* New York: Academic Press.

O'Riordan, Tim, and Jill Jäger. 1996. "The history of climate change science and politics." In *Politics of climate change: A European perspective,* edited by T. O'Riordan and J. Jäger. London: Routledge.

Rodhe, Henning, and Robert Charlson, eds. 1998. *The Legacy of Svante Arrhenius. Understanding the Greenhouse Effect.* Stockholm: Royal Swedish Academy of Sciences.

Schneider, Stephen H., and Randi Londer. 1984. *The Co-evolution of climate and life.* San Francisco: Sierra Club Books.

Stevens, William K. 1999. *The change in the weather. People, weather and the science of climate.* New York: Delacorte Press.

Web Sites

Contemporary work on global climate change, both scientific and political, is in rapid flux. The information to be found in most newspapers, news magazines, and television news is deeply inadequate. There are many good books. Houghton (above), perhaps the best overall short summary at the time of this writing, is already getting dated. To go deeper, there are reports available online issued by the IPCC, National Academy, and other institutions that work from scientific consensus. Some Web sites worth consulting at the time of this writing are:

http://www.ngdc.noaa.gov/paleo/
   NOAA's climatology site has tutorials on global warming and paleoclimatology, data, pictures, etc.
http://www.gcrio.org/
   U.S. government-sponsored global change research information site includes answers to basic questions
http://www.usgcrp.gov/
   U.S. interagency research program, including:
http://globalchange.gov/
   News items
http://www.usgcrp.gov/usgcrp/nacc/
   Reports assessing impacts

http://www.ipcc.ch/

    The IPCC's site, with its reports

http://www.nap.edu/

    National Academy Press, with many key reports

http://www.cnie.org/nle/crsreports/climate/

    Congressional Research Service reports

http://www.pewclimate.org/

    Pew Center on Climate Change, news and policy-related reports

http://www.wri.org/climate/

    World Resources Institute (mainstream environmentalism), with

http://www.wri.org/climate/climlinks.html

    A good links page

http://www.safeclimate.net/

    What you can do

http://www.globalwarming.org/

    Industry-funded site with arguments against the IPCC consensus

http://www.marshall.org/

    Marshall Institute's arguments against the IPCC consensus

http://www.greenpeace.org/~climate/

    Environmental activists, information and programs

http://www.environmentaldefense.org/

    Environmental activists, information and programs

http://climateark.org/links/

    Hundreds of links